# Solar Power Your Home

## FOR

# DUMMIES®

## 2ND EDITION

**by Rik DeGunther**

**WILEY**

Wiley Publishing, Inc.

**Solar Power Your Home For Dummies® 2nd Edition**

Published by
**Wiley Publishing, Inc.**
111 River St.
Hoboken, NJ 07030-5774
www.wiley.com

WILEY

# About the Author

**Rik DeGunther** attended the University of Illinois as an undergraduate and Stanford University as a graduate student, studying both applied physics and engineering economics (some of this education actually stuck!). Over the course of an engineering career, he worked as a project engineer, project manager, and product designer. He holds several United States patents and has designed a wide range of technical equipment, including solar power systems, military radar jammers, weather measurement equipment, high-powered vacuum tubes, computerized production hardware, golf practice devices, digital and analog electronic circuits, unmanned aerial vehicles, guitars and amplifiers, microwave systems, explosive devices (strictly by accident), cloud height sensors, fog sensors, furniture, houses, barns, rocket ships, dart throwers, flamethrowers, eavesdropping devices, escape routes, nefarious capers, and you name it. He's one of those nerdy guys who likes to take things apart to see how they work and then put them back together and try to figure out what the leftover parts are for.

Rik is CEO of Efficient Homes, an energy efficiency auditing firm in Northern California. He is actively engaged in designing and developing new solar equipment, including off-grid lighting systems and off-grid swimming pool heaters. He writes weekly op-ed columns for the *Mountain Democrat,* California's oldest and most venerable newspaper. He has also written a golf book (on putting) and spends most of his free time attempting to improve his relatively impressive but objectively droll golf handicap, usually to no avail. Sometimes the urge strikes him to play a very loud guitar, of which he owns a collection with far more intrinsic quality than the playing they receive. His hearing has been faltering the last few years, so he rebuilt his amplifier to go up to 11.

# Dedication

Of course, this book is dedicated to Katie, Erik, and Ally. Without them, the sun would not shine at all, and this work of art wouldn't exist.

# Author's Acknowledgments

Many thanks to all those who have contributed to the material in this book, whether wittingly or not: Dick and Betty DeGunther, Professor Mitchell Weissbluth, Professor A. J. Fedro, Professor Lamb, John Lennon, Paul McCartney, Leland Stanford, Mike Pearcy, Jordan Cobb, Eric Micko, Vikki Berenz, Connie Cowan, Betsy Sanders, Jim DeGunther, Sarah Nephew, Freddie Mercury, and Dave and Gary Romano of Sierra Valley Farms. Thanks to Robin Harp (Rik Manes and Reuben Veek) and the crew at Solarecity in Roseville, California, for educating me on the tricks of the solar trade. Thanks to Dr. Keith Kennedy and Watkins-Johnson Company for showing restraint above and beyond the call of duty. And thanks to John Steinbeck for making me understand what's important and what's not, and in the same vein, Derek Madsen.

Thanks to Mike Baker and Tracy Barr from Wiley and to the technical editor Greg Raffio for his excellent and well-received insights and for generously giving his time to this project. And thanks to Stephany Evans at Imprint Agency for getting all the ducks in a row.

Last but not least, thanks to all the *For Dummies* fans out there who have made the series what it is today.

## Publisher's Acknowledgments

We're proud of this book; please send us your comments at http://dummies.custhelp.com. For other comments, please contact our Customer Care Department within the U.S. at 877-762-2974, outside the U.S. at 317-572-3993, or fax 317-572-4002.

Some of the people who helped bring this book to market include the following:

*Acquisitions, Editorial, and Media Development*

**Project Editor:** Tracy L. Barr
  *(Previous Edition: Kelly Ewing)*

**Acquisitions Editor:** Mike Baker

**Copy Editors, Previous Edition:** Danielle Voirol, Vicki Adang

**Assistant Editor:** Erin Calligan Mooney

**Editorial Program Coordinator:** Joe Niesen

**Technical Editor:** Gregory Raffio

**Editorial Manager:** Jennifer Ehrlich

**Editorial Assistant:** Jennette ElNaggar

**Cover Photos:** © iStock

**Cartoons:** Rich Tennant
  (www.the5thwave.com)

*Composition Services*

**Project Coordinator:** Katherine Crocker

**Layout and Graphics:** Joyce Haughey, Christine Williams

**Proofreader:** Linda Seifert

**Indexer:** Potomac Indexing, LLC

**Publishing and Editorial for Consumer Dummies**

  **Diane Graves Steele,** Vice President and Publisher, Consumer Dummies

  **Kristin Ferguson-Wagstaffe,** Product Development Director, Consumer Dummies

  **Ensley Eikenburg,** Associate Publisher, Travel

  **Kelly Regan,** Editorial Director, Travel

**Publishing for Technology Dummies**

  **Andy Cummings,** Vice President and Publisher, Dummies Technology/General User

**Composition Services**

  **Debbie Stailey,** Director of Composition Services

# Contents at a Glance

# Table of Contents

# Introduction

● ● ● ● ● ● ● ● ● ● ● ● ● ● ● ● ● ● ● ● ● ● ● ● ● ● ● ● ● ● ● ● ● ● ● ● ● ● ● ● ● ● ● ● ● ● ● ●

*S*olar Power Your Home For Dummies, 2nd Edition, can help you turn your solar inspirations into reality without losing your shirt or your sanity. With this book — and with some good, hard work and perseverance on your part — you can achieve the laudable goal of installing solar power equipment at your home.

## About This Book

This book provides you with an ample solar knowledge base and lets you apply that info through do-it-yourself projects — or through hiring others to do the work for you. I use plain English to the extent possible, breaking down complex technical concepts into bite-sized pieces. But my goal is always to help you navigate the maze of solar technology as efficiently and sensibly as possible.

Every house is different. Climates are different, even in areas separated by a couple of miles. Markets are evolving, suppliers are changing, and technology is constantly evolving. Sometimes you're going to have to make your own decisions, and I can give you only the pertinent information you need to make the best one. Or in some cases, all I can do is point you toward the information sources you need to consult. But you can find the essentials here, tailor them to your own situation, and get a solar system that really works for you.

## Conventions Used in This Book

For simplicity's sake, this book follows a few conventions:

- ✔ *Italicized* terms are immediately followed by definitions.
- ✔ **Bold** indicates the action parts in numbered steps. It also emphasizes the keywords in a bulleted list.
- ✔ Web addresses show up in `monofont`.
- ✔ When this book was printed, some Web addresses may have needed to break across two lines of text. Rest assured that I haven't put in any extra characters (such as hyphens) to indicate the break. Just type in exactly what you see in this book, pretending as though the line break didn't exist.

# Foolish Assumptions

Here are some things I assume about you:

- ✔ **You want to add a solar component to your home.** I assume that you've already decided to move but aren't really sure which is the best direction or the best speed. You want to make the right decisions, and you understand those decisions are entirely yours to make.

- ✔ **You have some do-it-yourself skills.** You — or whoever's helping you install the equipment — can handle a screwdriver and power drill. You may know the basics of plumbing or electrical work.

- ✔ **You want to get things done as efficiently and sensibly as possible.** (Note I didn't say *quickly,* because that leads to errors in both judgment and facilitation.) You don't necessarily have an engineering degree, and you don't want to know every technical detail concerning the various solar technologies.

- ✔ **You need working knowledge of a project so that you can hire professionals and discuss matters with them.** You don't plan on handling a major project yourself, but you want enough information to make informed decisions.

- ✔ **You've got the right attitude for tackling projects.** Projects rarely, if ever, go the way you envision when you were drawing up the plans. As a lifelong practitioner of projects big and small, I've learned to be patient and enjoy the road every bit as much as the destination. When you run into problems — which you will — smile and take a step back and come up with a witty joke. The harder a project is, the more rewarding it'll be when it's finished. I can pretty much guarantee that you're going to find out precisely what I'm talking about.

# How This Book Is Organized

This book has seven parts. Without further ado, here are the parts of the book and what they contain.

## Part 1: Playing the Energy Game

In this part, you find info on doing a home energy audit so that you can understand exactly how and where you use energy in your home. I also give you a guide to making improvements in your house so your energy consumption's more efficient.

# Part II: Understanding Solar — Just the Facts, Ma'am

For most users, this part may be the most difficult to read because the technical and financial concepts can be elusive. Hang in there — understanding the nuts and bolts can make it easier to make the right decisions later on. After you get through the preliminaries, I help you decide which direction is best for you, and then you're on your way.

# Part III: Applications Aplenty: Projects from Small to Large

This part dives into solar power projects that you can start with right now. I describe a lot of interesting and useful little gadgets and small systems that can make your life easier and more fun. Regardless of how ambitious your solar aspirations may be, this is a good place to start. Greenhouses and solar rooms have some fundamentals that are consistent with all projects that you need to understand, so I cover the basics of what you need to know.

# Part IV: Exploring Full-Scale Photovoltaic Systems

Part IV gets into photovoltaic (PV) projects. I describe the different types of equipment that are on the market, and how best to choose the right equipment for your particular application. I give advice on deciding whether to do a project yourself or to hire a contractor, and I give advice on how to hire someone. I also detail how to specify a large-scale PV system and how to find and hire a contractor to do the installation for you. Financing is important because PV costs a lot; I give some useful advice on how to work with subsidies and rebates, and also banks and lending institutions. For those who want to go off-grid, I tell how best to do it and under what circumstances it's merited.

# Part V: Buying, Selling, and Building a Solar Home

In Part V, I give a broad range of advice on how to value a solar home, whether you're on the buying or selling end. And I tell you how best to find a bare lot and develop your dream of building your own solar home.

## Part VI: The Part of Tens

Like every *For Dummies* book, the Part of Tens includes quick resources that provide plenty of information and sage advice compacted into few words. Above all, this part demonstrates that you aren't alone. Gain wisdom from other solar energy enthusiasts' trials and errors.

# Icons Used in This Book

The icons in this book can quickly steer you to the information you need. Here's what they stand for:

The Tip indicates a paragraph that elaborates on a nifty little shortcut or frustration-saver.

This icon highlights important information to store in your brain for quick recall at a later time.

This icon should never be ignored. It points out things that can potentially harm you or your project. Remember: A successful project relies on many factors, both economic and practical, but the most important gauge of a successful project is that nobody gets hurt!

The Technical Stuff icon lets you know that some particularly nerdy, technoid information is coming up so you can skip it if you want. (On the other hand, if you want to read it, you don't actually have to be a nerd — just able to read.)

# Part I
# Playing the Energy Game

The 5th Wave                    By Rich Tennant

"Hold off on that. I think we're going to get solar panels."

## *In this part . . .*

You may be ready to go solar, but first you need to take a look at how you use energy in your household and how your energy bill is measured and calculated. Solar investments can save you money and make your life more comfortable, but part of the process is understanding exactly what types of solar investments you should make — and how big they should be. Here, I show you how to systematically dissect your energy bills and how to look around your house to find all kinds of energy conservation measures to reduce your energy bill even before you invest in solar.

# Chapter 1

# Helping the World through Solar Power

. . . . . . . . . . . . . . . . . . . . . . . . . . . . . . . . . . . . . . . . . . . . . . .

*In This Chapter*

▶ Investing in solar energy

▶ Looking at the benefits of solar power

▶ Solving for challenges

. . . . . . . . . . . . . . . . . . . . . . . . . . . . . . . . . . . . . . . . . . . . . . .

*P*roducing energy can be dirty work. Carbon emissions, coal slurry, nuclear waste, and other pollutants can wreak havoc on the environment, cause health problems, and make people hopping mad. And many energy sources are in limited supply, particularly fossil fuel sources that have traditionally dominated the world's energy usage. Not only does that drive prices up, but it also leads to political conflicts when people decide they're not willing to share. You're probably not ready to go completely unplugged, but you do want to play your humble part to save the environment, help the country become less dependent on foreign energy sources, and save money. Tall order? Maybe not. Above all the energy sources in use today, solar shows the most long-term promise for solving the world's energy problems. Solar power works well on both large and small scales, and it is possible to start using solar power right away. You can start small, and work your way up.

On any given day, 35,000 times the total amount of energy that humans use falls onto the face of the Earth from the sun. If people could just tap into a tiny fraction of what the sun is providing each day, society would be set. Of course, some problems do crop up, but they're solvable, and going solar can be well worth the effort.

To understand the role solar energy can play in your home, you need to have a good understanding of where your own energy comes from, where it's used, and how much pollution each of your energy sources generates. In this chapter, I explain how solar fits into your day-to-day life — and why it's such a good energy option.

# *Looking for Sustainable Energy*

The words *renewable* and *sustainable* are being knocked around quite a bit, and both are strongly associated with energy conservation. *Renewable* forms of energy constantly replenish themselves with little or no human effort. Solar energy is just one example — no matter how much you use, the supply will never end (okay, it may end after billions of years, but your using solar power won't make the sun burn out any faster). Other examples of renewables include firewood, water (through hydroelectric dams), and wind power. Note, however, that firewood is notoriously polluting; the term *renewable* does not necessarily imply good environmentalism. Firewood also has another potentially severe drawback in that people go out into forests and cut down trees, often without much thought to the overall health of the forest (a good example of not seeing the forest from the trees).

To make sure that resources last, humans need to focus on conservation, recycling, environmental restoration, and renewable and alternative energy sources. *Sustainability* is commonly associated with such a holistic approach to personal lifestyle. Not only are *sustainable* forms of energy renewable, but they also have the ability to keep the planet Earth's ecosystem up and running, in perpetuity. Sustainable energy, such as solar, is nonpolluting to the greatest extent possible. The basic notion behind sustainable energy sources is that by their use, society is not compromising future generations' health and well-being, nor their ability to use their own sustainable resources to any less capacity than we have in the past. Who can argue with this very fundamental version of the Golden Rule?

## Consuming the Earth

Here are some statistics about power use in the United States (from DOE):

✔ Americans import more than half their fossil fuels. Thirty years ago, this figure was only 33 percent, and analysts predict that within a few years, it'll rise to 66 percent. Even though new energy reserves are being found, our increasing consumption of energy is more than offsetting our increased domestic production.

✔ Of all the energy used in the United States, 39 percent comes from oil, 23 percent from natural gas, 24 percent from coal, 6 percent from hydropower dams, 7 percent from nuclear, and only 1 percent from renewables such as solar energy. On the plus side, the use of renewables is increasing much faster than other forms of energy, particularly with all the government subsidies and incentives that are being promoted.

✔ Americans get 51 percent of their electrical production from coal, 20 percent from nuclear, 18 percent from natural gas, 2 percent from petroleum, and only around 9 percent from renewables, of which the vast majority is hydro (water). Solar plus wind accounts for only around 0.18 percent of the grand total.

# Understanding Why Solar Is King

Solar power has historically been more expensive than other energy options, but that's changing fast because of government investment in technologies, as well as the simple fact that many more people are investing in solar, which results in economies of scale. Solar energy equipment increases your financial standing in basically three ways:

✔ Savings on your monthly utility bills.

✔ Appreciation of your home's value.

✔ Predictability in your utility bills for years to come. When utility rates increase, you'll be largely immune from the increases because you'll be getting your energy from a local source that's not tied in to the utility. Of all these three factors, this one seems to drive people to invest in solar the most.

The following sections cover reasons why solar is a great investment, both financially and environmentally.

## Reaping financial rewards

Solar is an investment; you must actively go out and purchase solar equipment and install it at your home. However, after the initial costs, not only do you save money from lowering your energy bill, but you will also see the value of your home increase.

### Comparing savings

So how does investing in solar compare to other investments, such as the stock market, a savings account, or a new kitchen?

To compare investments, you need to calculate your payback period. *Payback period* is a measure of how long it takes to recoup your upfront investment with the costs you save by installing solar equipment. If you install a solar water heater system for $4,000 and it saves you $50 a month on your power bill, the system will pay for itself in 80 months, or 6⅔ years. (Though you may easily cut that time in half if the price of oil skyrockets and utility rates double, for example, during a war in the Middle East.)

Now consider other ways you can spend that money. With investments in remodeling, such as a new kitchen, you get no monthly cost reductions at all unless you're installing new appliances that are more energy-efficient. Historical data indicates that if you remodel your home with a new kitchen, you'll only recoup around 70 percent of the cost of the remodel when you sell your home. If you put the same $4,000 into an interest-bearing bank account, you may get $20 a month in interest (that's at an interest rate of 6 percent, which is difficult to find these days). After 80 months, you'd make $2,000 in compounded interest, or half your investment. And if you put the same money into the stock market, you may enjoy a return of $3,400 in a single year. Of course, you can also lose the entire thing and drive yourself nuts with regret!

When you install a solar PV system, historical data indicates that you will recoup nearly 100 percent of your investment. If you spend $25,000 on a PV system, your home's value will increase by around $25,000, maybe even more if utility rates increase precipitously. It may be said that no other investment that you can make in your home will recoup as much.

To play it safe, choose a variety of investments and decide how much you want to put toward solar power. Stock portfolio managers consider hedging an important facet of a good portfolio. *Hedging* basically entails spreading the risk around over a range of individual investments. That way, if one of your investments goes sour, the effect on your entire investment portfolio will be minimal. "Don't put all your eggs in one basket," as the saying goes.

In short, if you install solar, you'll be relatively risk free from exploding energy costs. If you install a solar PV system that produces as much electricity as you use in your home, you'll never have to worry about paying another electric bill. And you won't have to worry about utility rate increases.

### Showing a little appreciation

When you go solar, your home *appreciates,* or increases in value. Realtors can give you statistics that estimate how much the value will go up, given the type of investment and the area you live in.

According to the National Association of Real Estate Appraisers (NAREA), for every dollar you save annually in energy costs with solar equipment, the value of your home increases by up to 20 times your annual energy savings, depending on the type of system you install. For a solar water heater investment of $4,000, the value of your home may increase by at least that much! How can this be? Solar is catching on, and homebuyers are willing to pay more for solar homes that promise energy savings. People are more and more willing to invest in energy systems that are clean and reliable, compared to the fossil fuel mainstay. In some communities, a solar home will sell much faster than a conventional home, and this may be important if you need to move quickly.

### Taking advantage of subsidies

Right now, a wide range of government and industry programs are available to help you finance your solar investments. Governments are giving out tax breaks, utilities are offering rebates, and low-interest loans are available for solar investments. The net effect is to make your solar projects less expensive and more attractive on the bottom line. With any solar investment, it's important to consider the net cost, which is the original cost of the equipment minus any subsidies and rebates. For a PV system, the net cost can be as low as one half of the original cost, which is a considerable savings.

In Chapter 20, I show you how to find the right resources for subsidies and rebates and tax credits, and how to work with the various agencies that you'll need to interact with.

## Erasing your carbon footprint

Most energy resources are burned in order to create the useable energy we so take for granted. The worst offender, in terms of pollution, is coal, and the United States gets around 50 percent of its electrical power from coal-fired power plants. Put simply, there's no clean way to burn coal, and that situation is not likely to change in the next decade, or even longer. While you hear a lot of talk these days about "clean burning coal" plants, it's all relative. A clean-burning coal fired power plant is simply cleaner than those that exist now; in the grand scheme of power generating systems, coal is one of the worst offenders in terms of greenhouse gas emissions.

Your carbon dioxide footprint is a measure of how much carbon dioxide you're releasing into the environment by virtue of your energy-consuming habits. A typical American *carbon dioxide footprint* is around 36,000 pounds (18 tons) per year. That's a lot!

Solar, however, has no carbon footprint, other than the energy is takes to manufacture a solar panel (referred to as *grey* or *invested* energy). For each kilowatt-hour (kWh) of energy-generating capacity you install with solar, you'll save that much from other sources, most likely the electrical power grid. No other alternative energy resource can offer this impact except wind power and hydro, but solar is far more versatile and widely available. Wind power is practical only where there's a lot of wind (which doesn't include all the hot air blowing around in Washington, DC). Hydropower is only available where suitable water resources are abundant, and hydropower disrupts the environment in ways that have resulted in the environmental movement frowning upon further hydropower development. (Plus all significant hydro sources in the U.S. have been tapped, anyway.)

Your carbon footprint is valuable for calculating cost versus gain for installing solar systems because — face it — even though pollution isn't costing you directly in your wallet, you need to factor it into your thinking.

When you generate solar electricity, you don't need transmission lines and all the associated inefficiency. Solar is right there, where you use it. When you install a 3kWh active solar system, you're offsetting the need for that much power from your utility company. But you're *saving* about 9kWh of total power consumption because of inefficiencies in the power grid. Therefore, you're actually saving much more than 3kWh, as well as the associated carbon footprint.

Table 1-1 can help you calculate your own carbon footprint.

| Table 1-1 | Carbon Emissions for Burnable Energy Sources | |
|---|---|---|
| *Type* | *Pounds $CO_2$/Unit* | *Unit* |
| Oil | 22.4 | Gallons |
| Natural gas | 12.1 | Therms (Btus) |
| Liquid propane | 12.7 | Gallons |
| Kerosene | 21.5 | Gallons |
| Gas | 19.6 | Gallons |
| Coal | 4,166 | Tons |
| Wood | 3,814 | Tons |

| Start with your car: | Your numbers | Example: |
|---|---|---|
| How many miles do you drive per year? | _____ | 15,000 miles |
| Mpg? | _____ | 23 miles per gallon |
| Divide to yield number of gallons/year | _____ | 652 gal/yr |
| Multiply by 19.6 (from above table) | _____ | 12,782 lbs/yr (ouch!) |

To find how much carbon dioxide you produce by using home fossil fuels, multiply the amount of fuel you use by the value in the second column, the pounds of $CO_2$ per unit. For example, suppose you use 400 gallons of home heating oil; you produce 8,960 pounds of carbon dioxide per year:

400 gal/yr × 22.4 lbs. $CO_2$/gal = 8,960 lbs. $CO_2$/yr.

And here's your carbon emissions for the 50 gallons of liquid propane you may use for your barbecue:

50 gal./yr × 12.7 lbs. $CO_2$/gal = 635 lbs. $CO_2$/yr

And here's the footprint for using 1 ton of firewood in a year:

$$1 \text{ ton/yr} \times 3{,}814 \text{ lbs. } CO_2/\text{ton} = 3{,}814 \text{ lbs. } CO_2/\text{yr}$$

Add those together, and your home fossil fuel consumption produces 13,409 pounds of carbon dioxide per year. Ouch! You can cut way down on that if you switch to solar heating and cooking, not to mention all the other solar options that are available.

Then calculate your carbon emissions from electricity use. This number depends on how your local power generators operate. Nuclear reactors emit very little carbon, and coal-fired generators emit quite a bit. The average North American value is 1.33 pounds of $CO_2$ per kWh. If you're using nuclear energy, you can reduce this number to about 1.0 pounds of $CO_2$ per kWh, or less. If you're strictly relying on coal-fired electricity, the number could go as high as 2.0. Here's how you calculate the carbon dioxide output if you use 10,000 kWh of energy:

$$10{,}000 \text{ kWh/yr} \times 1.33 \text{ lbs. } CO_2/\text{kWh} = 40{,}000 \text{ lbs. } CO_2/\text{yr (Youch!)}$$

## Enjoying solar's unlimited supply

At sea level, on a sunny, clear day, 1 kWh of sunlight energy is falling onto a 1-square-meter surface per hour. Over the course of a sunny day, you can realistically expect to capture around 6 kWh of total energy from this same surface area. That's 180 kWh per month. Five square meters is enough to completely replace a typical monthly power bill! If only it was so easy.

If you were to build an active solar panel measuring 100 miles by 100 miles in sunny Nevada (where you can get plenty of government land for free), you'd be able to produce enough power to handle all the United States' electrical requirements (except when it rained a lot!).

## Exercising your legal rights to sunlight

You have legal rights to your sunlight; nobody can build up so that your solar exposures are affected. Government acknowledges value in the amount of sunlight that hits your home.

You have a legal right to demand that your neighbors remove trees and other impediments to your solar access. If a neighbor's trees are shading your property, you can do something about it. Remember though, this right goes both ways. If you're shading somebody else's property, he or she can force you to remedy the situation. Check with your local governments to see what sorts of laws apply to your specifics.

## *Appreciating solar energy's versatility*

You can use solar energy in many ways, each with different costs and complexity. Later chapters and the upcoming section titled "Small to Supergiant: Choosing Your Level of Commitment" talk about some projects you can tackle. But for now, consider that solar power lets you do any of the following:

- **Generate electricity for general use:** You can install a solar electric generating system that allows you to reduce your electric bills to zero. This is one of the most popular solar applications on the market today, and the growth in solar powered electrical systems is over 25 percent per year. (See Part IV.)

- **Cook:** Using the sun and your vivid imagination, along with a few easy-to-build ovens and heaters, solar power can help you put dinner on the table. (See Chapter 9.)

- **Practice passive space heating:** The sun can heat your house by strategic use of blinds, awnings, sunrooms, and the like. (See Chapters 9, 13, and 15.)

- **Heat water:** Use solar energy to heat your domestic water supply — or let sun-warmed water heat your house by pumping it through appropriate plumbing systems. You may need no electrical pumps or moving parts other than the water itself. (See Chapters 10, 11, and 12.)

- **Pump water:** You can slowly pump water into a tank when the sun is shining and then get the water back anytime you want. You can also make your tank absorb sunlight and heat the water, thereby reducing the power load on your domestic water heater. (See Chapter 14.)

- **Heat your swimming pool:** You can cover your pool with a solar blanket to heat it cheaply and efficiently. Or you can install solar hot water heating panels on your roof that can heat your pool year round. (See Chapter 11.)

- **Add landscape lighting:** You can put small, inexpensive solar lights around your yard and eliminate the need for high-priced overhead lighting powered by the utility company. With advances in technology, these lights actually look and work better than hard-wired versions. This is the most widely accessible solar technology, and it's nearly fool proof. (See Chapter 8.)

- **Provide indoor lighting:** The technological boom in light-emitting diodes (LEDs) — small, electronic lights that take very little current and provide long lifetimes — has enabled a number of effective solar lighting systems for in-home use with very low power requirements. You can light your porches and even rooms in your house with a small, off-grid photovoltaic system connected to a battery. During the day, the battery charges so that you have enough juice at night to do the job. (See Chapter 9.)

- **Power remote dwellings:** You can completely power a remote cabin, RV, or boat with solar. (See Chapter 18.)

# Gaining independence from fossil fuel sources

In the United States, domestic supplies of fossil fuels are dwindling and demand cannot be met at the current rate of consumption growth. Even if new reserves exceed expectations and next-generation oil and gas recovery technologies significantly improve, supply and demand are going to be imbalanced.

In terms of dollars, energy imports accounted for around 24 percent of the U.S.'s $483 billion trade deficit in the year 2002 (note that coal is the only energy source that the United States doesn't import). Here's a breakdown of how much we import of the most common energy resources:

- Fossil fuels: 60 percent
- Natural gas: 16 percent
- Uranium used in nuclear reactors: 81 percent
- Total net fuel import: 33 percent

The U.S. imports more oil from Canada than any other foreign nation. Next in line is Saudi Arabia, then Venezuela, then Mexico. Each of these countries provides around 1.5 million barrels to U.S. markets every day. In particular, the Persian Gulf region supplies the U.S. with around 22 percent of its oil imports, or around 12 percent of total U.S. energy consumption. Our dependence on these sources of oil makes us beholden to the supplying nations, and they play political games with us, as a result. We simply cannot sustain the current level of energy imports, not only because the trade deficit suffers, but also because we are subjected to dubious political pressures from our chief import countries.

The Alaska Arctic National Wildlife Refuge (ANWR) has considerable oil reserves, but it is estimated that at the peak of drilling it will only be able to produce less than 1 million barrels per day. That's a mere drop in the bucket in relation to the amount of energy that we use.

Solar power, on the other hand, is domestically produced, and every kWh of solar energy that we produce reduces our demand for foreign oil sources by the same amount.

Electrical power is our most important form of energy consumption, and electrical consumption is increasing every year because electricity may be used in so many different ways. Solar power directly supplants the need to produce electrical power the majority of which comes from coal-burning power plants.

Some people believe that we need to promote electric vehicles, which use batteries to power the transmissions. These batteries need to be charged, and the current scenarios call for plugging in to the grid for the power. Doing so will result in a huge increase in grid-powered electrical demand, which in turn means that we'll be using a lot more coal and emitting a lot more carbon dioxide. An ideal solution is to use solar power, instead of the power grid, to charge the batteries. In some communities, special rate structures are available for those who charge their electric vehicles with solar panels. Look for this sort of thing to proliferate in the next decade.

## Eliminating peak power grid problems

In the summertime, on a hot day, air-conditioners run non-stop. The highest demands come in the late afternoon, when the heat of the day is most intense. Worst of all are weekdays, when businesses and offices provide their employees with a lot of conditioned air.

Utilities rely on huge power plants to provide their customers with the vast majority of their energy needs. It's a physical fact that the larger a generating plant is, the more efficiently it may be run (for more details, see my book *Alternative Energy For Dummies* [Wiley]). When peak power requirements exceed the capacity of these large, mainstay power generating plants, the utilities must obtain power from other sources, and these are generally inefficient and expensive, not to mention highly polluting.

In some cases, the utilities simply cannot provide the amount of power that their customers demand, in which case brownouts result (power is simply shut off). In California, brownouts were common in the early 2000s and resulted in people losing their power when they most needed it. Talk about getting hot under the collar!

Solar power systems generate their maximum outputs during the afternoons, when the sun is shining the brightest. Therefore, solar is a perfect solution to the peak power problems that are becoming more and more common across the country. In fact, the reason the state of California launched its solar subsidy program was to help mitigate the peak power problems. It wasn't out of concern for the environment, as most people believe. Solar is the perfect solution for peak power problems, and many utilities rely on their solar customers to help mitigate the need for peak power.

Utilities could, of course, solve the peak power problem by increasing their base capacity (the size of their main power plants), but this is extremely expensive and increases greenhouse gas emissions. The ideal solution to the peak power problem is to increase the use of solar electrical generating systems. By installing a solar generating system with battery backups, you'll be largely immune from power blackouts. Currently, many people install backup generators that run off propane or other fossil fuels just so they won't have to deal with power outages. A solar generating system provides the same backup capacity, with only a fraction of the air pollution. And you don't need an on-site tank for propane or fuel when you go solar.

# *Acknowledging the Dents in the Crown*

Sounds great! You're ready to go! But solar isn't all fun and games. The pros outweigh the cons — especially when you look at the big picture — but you should still understand the drawbacks. This section explains a few things to remember when working with solar energy.

## *Initial costs and falling prices*

Going solar requires an upfront expense. When you go solar, you get a good payback on your investment, but you do have to put out cash upfront. Most people don't want to bother, and many don't have the cash. There are a wide range of financing options (which I describe in Chapter 20), but financing can be difficult to obtain these days. Banks have become very selective; in general, you need equity in your home in order to qualify for a second mortgage, and many people have seen their equity disappear in during economic downturns.

Another issue to contend with is that the cost of solar varies quite a bit from year to year, so timing is an important concern. Buy now, or wait? Government subsidies play an important role in the net cost of solar equipment, and so politics plays a role in the equation. In the fall of 2008, for example, when the markets were plunging, the federal government increased the Investment Tax Credit from a cap of $2,000 to a straightforward 30 percent of the out-of-pocket price you pay after state rebates and other credits. This made a huge difference in the net cost of solar photovoltaics, and people who bought their systems prior to the change regretted not having waited for a few more months. Predicting how subsidies will change is impossible, but you must at least try to anticipate the future. A crystal ball may help, but there's no guarantee. (Unfortunately, there are no *For Dummies* books on predicting the future.)

## *Reliability and timing*

Solar works only when the sun is shining. If you want energy at night or on a dark day in the winter, you need either batteries or other energy resources. What makes sense in Arizona doesn't necessarily make sense in Seattle, Washington. Ultimately, solar relies on Mother Nature's generosity, and this varies from region to region. In fact, it even varies over different locations at your home.

Also, timing of energy use can make a lot of difference in how your utility bills add up. In a typical scenario, solar energy availability is at its peak when the household power demands are minimal. It's out of phase with need. This scenario isn't much of a problem with solar water heaters because they inherently store the collected solar energy for later use. But solar electric requires either batteries for energy storage or a special system called an intertie, which connects to your public utility. In Chapter 16, I explain the technical details of an intertie system.

It's important to understand the affect that intertie has on your power bills. Figure 1-1 shows the typical residential energy requirements and the energy production of a solar system.

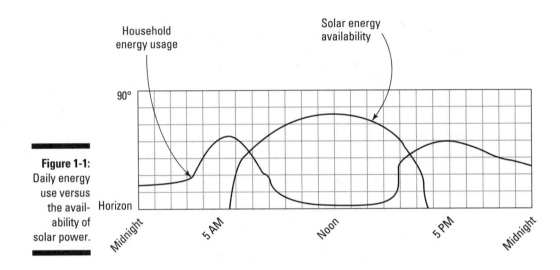

**Figure 1-1:**
Daily energy use versus the availability of solar power.

On this typical winter day, the heater is on all night but turned low, and the lights and appliances are off. In the morning, the family turns up the heater, turns on lights, takes hot showers, cooks breakfast, and gets ready for school and work. Then everybody leaves, and the day warms up so the heater shuts down. At the end of the afternoon, when the sun is on its way down, everybody returns. Lights are turned on, the heater's turned up, a log is tossed into the fireplace, cooking begins in earnest, the kids play video games and make a big mess, the vacuum is run, and so on.

To take full advantage of solar power, you may have to alter your energy consumption habits. Throughout this book, I explain how to make the best use of solar power. You can certainly install solar without changing your consumption habits, but you'll get a better return on investment when you understand how best to take advantage of your new technology.

## *Red tape and aesthetically minded neighbors*

You may have to work around building codes. Bureaucrats are a big hassle, and interfacing with government agencies is frustrating. In addition, only qualified contractors should install complex electrical systems. Even if you choose to install a complex system yourself (which you can usually do if you're well versed with tools), you have to deal with the county building codes and know how to schedule inspectors and handle the problems that may occur. Solar can be very simple and straightforward, but it can also entail a lot of technical details.

Also, most solar panels are ugly. Nobody wants to look at them. If they're your own and you're benefiting, it's acceptable. If they belong to your neighbors, it's a different story. In some communities, solar panels are forbidden. Many CC&Rs (Covenants, Conditions and Restrictions) prohibit solar panels altogether, but this situation is changing; in fact, most legal bans of solar panels are being stricken by the courts in favor of environmental conscience. At some point, the federal government will likely enter the picture and prohibit all banning of solar panels. Several ongoing efforts are working to make solar panels less obtrusive, so this problem will become less important over time. You can now buy solar panels that blend right in to your roof; it's almost impossible to tell the panels are even there. You can also buy panels in different colors.

## Effort and upkeep

Going solar takes work. Making good decisions about solar power can be difficult unless you've done your homework. And not only do you have to do some research, but you also have to work with the equipment itself. Here are some issues to consider:

- ✔ **You face some dangers.** Active electrical systems can shock you if you don't know what you're doing. Water heating systems can scald you. You're much safer sitting in front of your TV than climbing around installing solar equipment on your roof.

- ✔ **You face equipment challenges in freezing weather.** Solar water heating panels can freeze up in the winter. You have to pay attention to how they're working. Many new solar thermal heating systems get around the freezing problem by using some form of anti-freeze, but there are still a good number of existing and new systems that still use water exclusively.

    The anti-freeze systems are more expensive, but not everyone needs one. Be wary of contractors who are more interested in selling you the most expensive system possible than selling you the right system for your needs. It's ultimately up to you to do your homework and decide which system is the best for your application.

- ✔ **You're on your own for upkeep and repairs.** If you get your power from the power company, keeping things maintained and running is its problem. If you have a big array of solar panels on your roof, it's *your* problem. If they break, you pay. When they get old, you update. Warranties run for 25 years for solar PV panels, but you may have to pay some labor costs to have warranty work done. At the very least, you have to understand your system so that you'll know when it's not working properly. If a single panel goes out in a solar PV system, for example, the production may suffer to the tune of 25 percent or more. You're the one responsible for determining how your system is working and taking the necessary measures when it's not working properly.

✔ Solar panels affect other roof maintenance tasks. If you need a new roof, for example, you have to either pay a contractor to remove and then reinstall the panels once the new roof is in place, or you have to do the work yourself. In general, you can expect to pay a contractor around $1,500 to remove and reinstall panels when you put on a new roof. This estimated price fluctuates quite a bit, depending on the type of roof you have, and how large a solar system you're working with. In addition, while your panels are removed and the roof is being changed, you're not going to be generating any solar power. If you're saving $400 per month on your utility bills with your solar system, and it takes two months to install a new roof, you need to add $800 to the net cost of the roof job (in addition to the cost of removing and reinstalling the panels).

Consider the condition of your roof before you install solar panels. If you're going to need a new roof in a few years, you may be better off waiting until the new roof is in place before you purchase your solar system. If you're going to be changing your roof, ask your solar contractor for a bid to remove and reinstall the panels; get the contractor to guarantee the bid for a few years, if possible.

It's to contractors advantage to downplay the cost of removing and reinstalling panels. So watch out for unscrupulous contractors (more on them in the next section) who tell you that they'll move the panels for a very low price. Once you've got the system in place you're going to be stuck with whatever price they charge.

## Unscrupulous contractors and wild claims

Alternative energy is a hot topic these days, and you hear a lot of wild claims being propagated by the media, which sometimes fail to verify the accuracy of these claims, and by folks out to make a quick buck by selling something that's too good to be true. Some claims are obvious nonsense, like the story of the inventor who designed a new engine that gets 100 miles per gallon by using water as fuel. Others are a lot more difficult to judge.

Here's an example. A solar PV contractor in the Sacramento area where I live tells customers that he can install a revolutionary new energy saving device in the fuse box, so that the household uses half the power to achieve the same results. He offers to sell a solar system half the size of other contractors, since that's all that's needed, in conjunction with the revolutionary new device. Sounds great, doesn't it? But it doesn't work, and when the customer comes to realize that the wild claims are bunk, it's too late. You can always launch a lawsuit, but that's expensive and time consuming and your contractor may have already gone out of business because he's already been sued any number of times.

The solar power industry is changing daily, along with the technology. Following are suggestions on how to protect yourself:

- ✔ **Don't look just at the price; look at the system's long-term reliability.** It's almost always the case that time-tested technology works better than radical new technology. It's not just a question of whether something works or not; it's also a question of whether something works for years and years, because that's what you're ultimately looking for. Bottom line: In general, if it sounds to good to be true, it is. If you talk to a contractor who wants to install a radical new technology that will solve all your problems for half the cost that other contractors are bidding, run for the hills.

- ✔ **Pay attention to the amount of experience the contractor has.** With the economy suffering, there are a lot of "solar contractors" who were electrical contractors or plumbers only a year ago. They work out of their garage, and have very little overhead rate. You may get a fine system installed by one of these contractors, but you should wonder if your warranty will be honored a few years down the road. You should also wonder if the system will operate as optimally as possible, because experience counts for a lot. Every solar installation is a little different, despite the fact that the equipment may be exactly the same. It's in the details where good contractors pay off. You may be tempted by the lowest price, but you should be skeptical of the lowest price as well. You can get information on the veracity of contractors over the Internet.

- ✔ **Get conversant about solar power.** The simple fact is, you need to understand solar to the point where you can critically judge the equipment that you're buying. The less you need to trust the contractor's performance claims, the better off you will be. In fact, when you understand how your equipment is going to work, you can tell how good a contractor is by simply noting whether or not he exaggerates the future system's performance.

- ✔ **Ask for a performance guarantee.** Even with a good contractor, you can't always get this because performance depends on how much sunlight you ultimately get. But when you ask, you'll get a sense of your contractor's honesty by the way they respond to your request. In fact, if your contractor is more than happy to give you a performance guarantee, you should be skeptical.

# Small to Supergiant: Choosing Your Level of Commitment

No matter where you start, you can always expand your solar system. For example, you can invest in a small photovoltaic system for your rooftop and then expand it as you go, spreading the investment costs over a long period of time. Read on for ways to get involved.

## Small- to mid-sized projects

In Part III, I detail a wide range of simple, straightforward solar projects available to the do-it-yourselfers. From installing landscape lighting to a stand-alone photovoltaic powered attic vent fan, you can begin investing in solar today with minimal cost and effort. The small-scale projects feature safe operating levels (typical voltages in a photovoltaic system are so low that you won't be able to get a shock). And if you're no good with tools, have no fear. Some of the projects don't even require a screwdriver — you can buy off-the-shelf solutions that you can use out of the box.

You can also do a number of things in your yard to improve the solar exposure of your home. Deciduous trees (which shed their leaves in the fall) planted strategically about your house can ensure summertime cooling while allowing solar energy to help warm your house in the winter. Planting bushes in the right spots can reduce the cooling effect of wind, especially around your pool. And you can also increase the breezes flowing through your house by strategically arranging trees and bushes. Chapter 8 shows you how.

In Chapter 9, I describe a wide range of small-sized solar investments you can enjoy right now. Some of these projects, such as the attic vent fan, can save you money. It can keep your attic cooler in the summer and lower your air-conditioning bill. And some of the projects — such as the swimming pool light ball — are just plain fun. You can even build a solar oven that'll cook almost everything you cook in your kitchen stove!

Chapters 10, 11, and 12 explain how you can use the sun to heat water and how you can put it to work in your home. Installing a solar heating system on your swimming pool is a reasonable do-it-yourself project with very real results, not just in terms of costs saved but also added enjoyment. Installing a water heating system for your domestic supply gets a little trickier, and it's a project usually contracted out to the pros. But if you're good with tools, you can probably install one on your own.

If you're intending to install a full-scale solar energy system in your home, a great way to find out about the character of solar energy is to start with the simpler projects. You discover the importance of good solar exposure, and you determine when and how solar works the best — as well as the worst. You'll be in a better position to make good decisions on how to invest the big bucks when the time comes.

## Large projects

In Part IV, I describe the larger projects that are wise choices for solar invest-ment. Full-scale photovoltaic energy generators are the king of the mountain these days. You probably won't want to install one of these systems on your own, but you can, if you have the necessary skill. I tell you how to research the available options and decide which is best for your needs. And I tell you how to find and choose the right contractor for the job.

Greenhouses are attractive, and you can grow your own food in them, year-round in some climates. But you can also use a greenhouse or sunroom to heat your home in the winter. In Chapter 15, I show some examples of the most popular greenhouse projects.

*Off-grid living* means there's no utility company power coming into your home at all. You can use a solar power system, backed up with a gasoline-powered generator, to provide all the power you'll ever need. In Chapter 18, I describe the things you need to keep in mind if you decide to take this plunge. It's not for everybody, and it really doesn't make much sense unless you're living so far away from the utilities that just running the lines to your house would cost a ton. But for some of the more independent-minded readers, it's the only way to go.

## Designing a solar home from scratch

Designing a solar home from scratch is clearly the most efficient way to achieve solar energy advantages. Most existing homes are inefficient in a number of ways. Insulation may be lacking. Sunlight exposure was not thought out — it's just what happened when the house was built on the lot. But when you design your own home, you can control all the variables. You can achieve excellent sunlight in the morning while blocking off the after-noon heat. You can shelter for wind by taking advantage of existing trees and cover. Best of all, you can build your roof to achieve perfect solar exposure.

You can also ensure energy efficiency by using the right materials and building techniques. The fact is, a good house design can make it so that you don't need much energy at all. What could be better for the environment?

# Chapter 2

# Surveying Your Household Energy Efficiency and Use

*I* love moving forward and getting things done because I'm impatient and I want to see some results. I like pounding nails and gluing stuff together (although I don't like pounding my thumb, or gluing my pants to the workbench). But before you start the fun stuff, you need to do a little research first. Why? Investing in solar energy is expensive.

It's much cheaper, marginally, to invest in energy conservation measures for your home before you invest in solar. Furthermore, most people don't have a good handle on where and how they're using energy. When you invest in solar power, you're called on to decide, very specifically, what type of system to install and how much it's worth in both monetary and labor costs (labor costs relate to both installing the equipment and maintaining it once the system's in). By studying the details of your own household and how you consume energy, you'll be well along the way to making wise solar investment decisions.

 I use a friend of mine as an example in this chapter. His fictitious name is Bill Toomuch. He and his small family live on a 5-acre ranch, all electric, in a 2,700-square-foot house. There's an above-ground swimming pool and a hot tub. Water is supplied by a well. Analyzing Bill Toomuch demonstrates how to go about systematically analyzing your own situation. (Plus, it's fun to spy on other people and get into their private details.)

# Dissecting Your Energy Bills

How, when, and where do you use energy? And why? Looking at nationwide statistics will give you a baseline to compare your energy usage to. Does your own energy consumption make sense, or can you make wise and cost effective changes? Should you make changes? Most people are surprised when they find out exactly how much and where they're spending their hard-earned cash on energy. And most people find that they can easily save over 10 percent of their energy bills by simply making some easy changes that don't entail a reduction in their quality of life. The fact is, energy has been a relatively inexpensive commodity, and we tend to take it for granted. That's changing due to both higher energy costs and a broader concern for the environment.

As the saying goes, "The devil is in the details." In this section, I present a detailed system for analyzing your energy use.

## Collecting the data

The first thing you need to do is collect a stack of power bills if you don't already have them. Call your utility company to see whether the folks there will send you copies of your past power bills going back a few years. Most utilities don't charge for this service, and some may even do data analysis for you. Ask when you call. The company may even offer other services, such as a free home energy audit. In many parts of the country, utilities are required by law to provide energy conservation services; take advantage, if you can.

Some of you may be lucky enough to get your utility to graphically plot your data for you, as well as give you some handy analysis pointers. Ask whether this service is available.

You also need to collect records of all expenses even remotely related to your energy consumption. If you don't have records, re-create them as best you can. Don't worry — re-creating can serve your purposes just fine. The most important point is to be complete; leaving something out is much worse than approximating the data for items that you do include. Here's a sampling of what you may want on this list:

✔ **Wood and fuel:** List anything you burn for heat, light, or cooking — firewood, manufactured logs (such as Duraflame), wood pellets, bio-energy products, charcoal, propane, gasoline for lawn tools or a generator, kerosene for lanterns or cooking or heating, and so on. You'll probably be surprised at all the energy sources that you haven't even considered, but they all count in the grand scheme of things.

✔ **Collection and transportation:** Include the costs of gathering and transporting any fuels. Don't forget costs of gas cans; chainsaws, saw blades, or wood splitters for firewood; or propane tanks for the grill.

✔ **Rental fees:** Include rental fees for your propane tank, power generator, or any other items you don't actually own.

✔ **Equipment purchases:** Collect cost information for any new barbecues, water heaters, space heaters, dehumidifiers, portable air conditioners, swamp coolers (evaporative coolers), general HVAC systems, power generators, and so on.

✔ **Repairs and maintenance:** Note what you spend on maintaining and repairing equipment such as woodstoves. Figure out how much HVAC filters cost you.

✔ **Household items:** Estimate how much you spend on items such as candles and batteries.

## Divvying up costs, month by month

Allocating all your energy costs on a month-by-month basis is usually a straightforward process. Your best bet is to take several years' worth of data and find the average for each month. In other words, add the last two or three years' worth of February data and then divide by two or three to get your monthly average. In this way, you'll average out seasonal weather variations and other changes, such as vacations.

Keep in mind that you don't need to split hairs in this analysis. Just by analyzing your usage in a systematic way, you'll come to understand your energy picture much better. It's the structure of the analysis that's more important than the details. Just the act of doing it is more important than doing it with extreme precision.

*Note:* Many utilities do what's called *averaging,* where they don't actually read your meter every month. They come out periodically and read it and then average it over a few months. If this is the case, you can still get a good understanding of your energy habits, but some of the detail is lost. If you feel the need, call the company and ask how your bill is structured.

For other items, figuring out monthly costs can be more challenging. You want to divide up costs so that they reflect how much you use a certain energy source in each month. For example, if you have a propane tank that's filled periodically by the propane company, you may have a tough time figuring out the monthly usage. Just get the yearly total and divide that up on a monthly basis. Your estimates will be very good. If you use the propane only for heat, allocate the cost to those months in which you use heat. If you use some for hot water and some for heat, divide it up as best you can. Make sure that the yearly total comes out right. Ditto with firewood and all the associated expenses. Most people buy a big stack once a year. How much did you use each month? Divvy up the costs.

Servicing for equipment should likewise be allocated (by month) according to the use of that equipment. For example, if you service your heater in July for $240, it only makes sense to assign that $240 to those months when you use heat. If you service only once every three years, allocate the costs accordingly. Once again, don't worry if you're not perfect. Just do the best you can and you'll see that it's good enough.

For large equipment expenditures, accountants use a term called *depreciation*. Here's how it works:

1. **Determine initial costs.**

   For example, perhaps a gas fireplace costs $4,000 plus 7 percent sales tax, with another $200 for permits and inspection. The total is $4,480.

2. **Estimate how much that equipment increased the value of your home.**

   A gas fireplace may increase the value by $2,500. The best way to find out how much an investment affects the value of your home is to consult with a realtor or an appraiser. But you'll probably find that your own estimate is close enough, although most people are guilty of great generosity when it comes to this calculation. Just be reasonable.

3. **Subtract the increase in value from the initial costs to get the net cost of the equipment.**

   The net cost of the fireplace is $4,480 – $2,500, or $1,980.

4. **Estimate the lifetime of the equipment in terms of months.**

   The fireplace should be usable for 10 years, or 120 months. When you buy the fireplace, ask the store for a number and then multiply this amount by about 80 percent to account for generosity. (The store does want to sell you a fireplace, after all.) Another way you can estimate this figure is to simply use the lifetime of the equipment you're replacing as a basis. If you're replacing a 10-year-old fireplace, use 10 years for the estimated cost of the one you're selling.

**5. Divide the net cost by the lifetime of the equipment to calculate the monthly depreciation.**

Dividing $1,980 by 120 tells you that the fireplace costs $16.50 per month.

For an even more accurate picture, allocate the cost of the fireplace to only the winter months. For example, you use the fireplace only over the six months from October through March. Dividing the yearly cost ($198) by 6 tells you that the fireplace costs $33 per month in October through March and $0 per month in April through September.

## Checking out a sample electric bill

The best way to illustrate how to do an energy survey is to use Bill Toomuch as an example. Figure 2-1 shows a typical sample from Bill Toomuch's utility bills. Here's a list of some of his energy costs:

- His utility company is PG&E, in Northern California.

- He uses 3 gallons of gasoline a month in lawnmowers, leaf blowers, and so on. That's around $8 per month, year round.

- He uses $15 per month in propane for his barbecue, and because he's in sunny California, the usage is consistent over the course of the year.

- His HVAC system broke down, and it cost $267 to fix. A repair this size is typically done once every two years, so the monthly cost is around $11.

- In the winter months, he burned manufactured logs in an old fireplace and estimated the costs as follows: January $62, February $60, March $35, April $10, October $10, November $62, and December $96 (Christmas cheer!).

- Bill's wife loves candles; she spends about $20 per month on them.

- He spends $23 per month in batteries for remotes, flashlights, sprinkler controllers, and so on.

| Telephone Assistance |
|---|
| 1-800-743-5000 |
| Assistance is available by telephone 24 hours per day, 7 days per week. |

| Local Office Address |
|---|
| 4636 MISSOURI FLAT RD PLACERVILLE CA 95667 |

| Account Number |
|---|

February 2007

### ACCOUNT SUMMARY

| Service | Service Dates | Service |
|---|---|---|
| Electric | 01/12/2007 To 02/12/2007 | $402.79 |
| Energy Commission Tax | | 0.51 |
| | | |
| TOTAL CURRENT CHARGES | | $403.30 |
| Previous Balance | | 560.58 |
| 01/23 Payment - Thank You | | 560.58- |

| **TOTAL AMOUNT DUE** | **$403.30** |
|---|---|
| **DUE DATE - 03/05/2007** | |

Rate Schedule : E1 SH Residential Service
Billing Days :    32 days

| Serial | Rotating Outage Blk | Meter # | Prior Meter Read | Current Meter Read | Difference | Meter Constant | Usage |
|---|---|---|---|---|---|---|---|
| N | | J78867 | 45,435 | 47,746 | 2,311 | 1 | 2,311 kWh |

Charges

01/12/2007 - 02/12/2007

| Electric Charges | | | | $402.79 |
|---|---|---|---|---|
| Baseline Quantity | 1,033.60000 Kwh | | | |
| Balance Usage | 1,033.60000 Kwh | @ | $0.11430 | |
| 101-130% of Baseline | 310.08000 Kwh | @ | $0.12989 | |
| 131-200% of Baseline | 723.52000 Kwh | @ | $0.22944 | |
| 201-300% of Baseline | 843.80000 Kwh | @ | $0.32146 | |
| Net Charges | | | | $402.79 |

The net charges shown above include the following component(s).
Please see definitions on Page 2 of the bill.

| Generation | $190.05 |
|---|---|
| Transmission | 18.09 |
| Distribution | 142.04 |
| Public Purpose Programs | 18.38 |
| Nuclear Decommissioning | 0.69 |
| Trust Transfer Amount (TTA) | 14.60 |
| DWR Bond Charge | 10.84 |
| Ongoing CTC | 0.32 |
| Energy Cost Recovery Amount | 7.78 |

Taxes

| Energy Commission Tax | $0.51 |
|---|---|

| **TOTAL CHARGES** | **$403.30** |
|---|---|

| Usage Comparison | Days Billed | Kwh Billed | Kwh per Day |
|---|---|---|---|
| This Year | 32 | 2,311 | 72.2 |
| Last Year | 32 | 2,449 | 76.5 |

**Figure 2-1:** Sample electric utility bill.

Bill's home is all electric, so that makes up the lion's share of his total energy cost. Bill called his utility company and got four years' worth of history, broken down by month.

The billing date is from 1/12/2007 through 2/12/2007. The total for the bill is $403.30. (Kilowatt-hours) of usage is 2,311. (Just for interest, the average daily cost is $13.) Adding the expenditures from the miscellaneous items I list earlier, Bill's total energy bill for this period is $547. Ouch! Somebody's not turning the lights off when they leave the room (Bill has two kids).

On your own power bill, note your rate structure. For most people, this step is easy. The cost per kWh stays the same, no matter how much power they use. Some people, however, are billed on a tiered system, which basically means that their rates go up the more energy they use. Other rate structures may include time-of-use (TOU), which means you pay more during peak hours.

In Bill Toomuch's case, there's a baseline usage rate for the first 530.4 kWh at $0.11430. (I didn't even know a thousandth of a penny existed!) For 101 to 130 percent of that baseline (up to 689.52 kWh), the rate goes up to $0.12989 — not much of an increase. From 131 to 200 percent of baseline usage (up to 1,060 kWh) costs $0.22944, and 201 to 300 percent (up to 1,591.2 kWh) costs $0.32146 per kWh. This latter rate is nearly three times the base rate — quite a penalty. Usage over 300 percent is even more punitive, and so on.

### Finding costs per kWh

Electric bills often list the cost per kWh, but when you factor in all the extra fees power companies charge you, the price per kWh jumps. Check out the breakdown that energy companies provide. Generation, transmission, and distribution are pretty straightforward. But if your power company is anything like Bill's, you may be coughing up cash for public purpose programs, nuclear decommissioning, trust transfer amounts (TTA), DWR bond charges, ongoing CTC, energy cost recovery amounts, and the like. I'm an energy expert, and I don't know what any of this stuff means. Somehow I get the impression that the utility company prefers it that way.

To figure out how much you're actually spending, simply divide the total of your bill by how many kWh's you used. In Bill's case, this yields 17.5 cents per kWh. Comparing this amount to some other parts of the country is instructive. In central Florida, the overall rate is around 11.6 cents per kWh. In Milwaukee, it's around 10.4 cents. (As usual, California is leading the way into the future, which in this case is higher energy prices.)

To see how your rate compares, visit the Energy Information Administration Web site at www.eia.doe.gov. There, you can find a number of reports, including information on the retail prices of electricity on a state-by-state basis. Why is there such a wide variance between regions? Because government regulates the utilities, and government never makes any sense. There are also major differences in the way a utility generates power, and finances their power plants. For instance, nuclear generated electricity is often cheaper than coal plants. And the quality of utility companies varies; some are well managed, some are poorly managed, and this will eventually affect their prices. So don't try to make sense of any of it. If that sounds cynical, keep in mind that cynicism is just an unpleasant way of spelling out the truth.

### Interpreting the data

After you compile all your data, add to your monthly utility bills all the other energy costs to get a grand total for each month. The most important other costs are gas expenses such as propane, natural gas, liquid gas, and so on. Place your data in a chart.

Table 2-1 shows Bill Toomuch's monthly expenditures for the entire 2007 billing year.

| Table 2-1 | | Example of Annual Energy Expenditures Plotted by Month | | |
|---|---|---|---|---|
| **Month** | **kWh Used** | **Electricity Charges** | **Price/kWh** | **Total Energy Cost** |
| Jan. | 2,290 | $380 | $0.166 | $530 |
| Feb. | 2,449 | $433 | $0.177 | $581 |
| Mar. | 2,452 | $456 | $0.186 | $579 |
| Apr. | 2,266 | $407 | $0.18 | $505 |
| May | 1,715 | $290 | $0.17 | $378 |
| June | 1,676 | $325 | $0.194 | $413 |
| July | 1,948 | $404 | $0.21 | $492 |
| Aug. | 1,963 | $443 | $0.226 | $531 |
| Sep. | 1,444 | $267 | $0.185 | $355 |
| Oct. | 1,460 | $267 | $0.183 | $365 |
| Nov. | 1,663 | $312 | $0.188 | $462 |
| Dec. | 2,319 | $406 | $0.175 | $590 |
| **Total** | **23,645 kWh** | **$4,390** | **$0.186 (average)** | **$5,781** |

Here's what you can glean from the data:

✔ **Changes in energy usage:** The shape of the curve is typical. The heater is on a lot in the winter, and the air conditioner runs in the summer. If you use an air conditioner, your potential graph should look similar in shape, unless you have unusual climate conditions. Or perhaps you're more tolerant of heat or cold than most people.

- **Baseline usage:** You can establish your *baseline usage* by looking at the months in which no heating or air-conditioning are used at all. May and September are the usual candidates. In the example, Bill Toomuch's baseline usage is approximately 1,500 kWh. (Baseline usage is the energy consumption with both heating and air-conditioning factored out. This figure is of interest because it tells you how much you're spending on heating and air conditioning, versus the rest of your requirements.)

- **Average kWh per day:** Add up the total number of kilowatts you used that year and divide by 365 (or 366 in a leap year). The average American kWh per day is around 20, so see how your usage compares. Bill Toomuch uses over three times that. His house is bigger, with a lot of extra goodies, but he has plenty of room for improvement. I plan on giving him a stern lecture.

- **Seasonal rate changes:** Check your bills to find out whether you have a summer rate increase in relation to the winter months when the same amount of total power was used. This jump is common; the power company charges more during peak seasons (I provide more details on this subject in my book *Alternative Energy For Dummies*, Wiley).

## Accounting for anomalies

Not every drop or rise in energy usage reflects changes in your energy habits. Individual years are subject to strange weather patterns or maybe one-time-only events, such as a big wedding. Look at several years' worth of data and if you see any numbers that look out of place, try to figure out why. You may want to plot several years' worth of graphs to get a more comprehensive idea about your energy consumption.

Watch out for isolated events on your bills. For example, perhaps one of Bill Toomuch's monthly statements included a one-time deduction as a result of the power company's settling a lawsuit with energy providers who charged too much. You need to factor this sort of thing out because it has no bearing on your habits.

### Getting the time of day

You may also be on what's a *time of use* (TOU) billing system. This system requires a special power meter that can distinguish between how much energy you use during different portions of the day, such as from noon to 6 p.m. The power company bills you more for energy used during this peak time slot. You may have three or four different time slots, each with its own rate. If you have this type of system in place, you can create even more data lines on your graphs that'll both illustrate your habits and the costs of supporting those habits.

Do you have a really high bill one month? One that sticks out like a sore thumb? Here are the biggest reasons:

- The addition of occupants in your home

- The addition of appliances, like a freezer in the garage

- Faulty appliances

- The use of too much heating or air conditioning

- Use of appliances with large motors (pumps, compressors, air conditioners)

- Seasonal appliances — electric blankets, dehumidifiers, lots of shop lighting, power tools

- An estimated bill, or one that reflects a longer time period

- A wedding or a big party

- Uncle Bill's motor home parked in the driveway, with an extension cord running into your garage electrical outlet

# Adding Up Typical Energy Usage

You need to look at the details of how you're using energy at your house in order to determine the best way to spend time and money reducing your power bills — and how best to invest in solar equipment. A good start is to first look at how the average household in America uses energy. You can compare your own situation and get some real insights.

According the Department of Energy (DOE), the average household expends the following percentages on its energy consumption:

- 44 percent on heating and cooling

- 30 percent for lighting, cooking, and appliances

- 18 percent for water heating

- 8 percent for refrigerators

The single most expensive power requirements in any home are the heating, ventilation, and air-conditioning system, or HVAC. Answer these questions:

- What type do you have — wood burning stove, gas fireplace, electric heat pump, or something else? How do you pay for the energy in each? How old is the equipment? Are you spending a lot for repairs?

- How much would upgrading to a more efficient system cost? Assign dollar values to the alternatives. Most houses that have wood stoves also have other heaters as well. What's the tradeoff?

✔ If pollution is an important consideration to you, what's your carbon footprint for each alternative? If you have a wood stove, your cost may be very low (especially if you cut your own wood), but your carbon footprint may be astronomical. Leaky old fireplaces are notoriously inefficient, and they emit a lot of pollution to boot.

✔ Add all the separate costs and look at the grand total. Do this for several years. What's the trend?

✔ Have unit costs changed over the last few years? Natural gas, for example, fluctuates a lot in price. What's the trend? How does this affect your household economy?

Next, you want to analyze the power usage that's contained in your baseline. Table 2-2 lists some typical power consumption numbers for home appliances.

| Table 2-2 | Annual kWh of Usage for Various Appliances | | |
|---|---|---|---|
| *Appliance* | *Energy Usage (kWh)* | *Appliance* | *Energy Usage (kWh)* |
| Spa (pump and heater) | 2,230 | Dishwasher | 600 |
| Pool pump | 1,430 | Aquarium/ terrarium | 570 |
| Refrigerator | 1,200 | Well water pump | 500 |
| Washing machine | 900 | Dehumidifier | 357 |
| Clothes dryer | 845 | Microwave oven | 150 |
| Waterbed heater | 850 | Television | 140 |
| Freezer | 750 | Home computer | 107 |
| Electric cooking | 680 | Electric blanket | 98 |

Use Table 2-2 to estimate your own approximate usage. Modify according to your personal situation. For example, you can lower your estimates if you turn off lights religiously and use fluorescents exclusively. On the other hand, you have to raise the amounts if you

✔ Have kids, because they often need a lot more washing and drying of clothes.

✔ Take a lot of hot and/or long showers.

✔ Own a roomful of aquariums.

✔ Have a huge television with a powerful sound system.

✔ Charge up your golf cart in the garage every night — that costs a lot.

 ✔ Plug in your RV.

 ✔ Insist on lighting your yard all night long.

Spend some time going around your house, looking at all the different ways that you consume energy. You may come across things you never even thought of before, such as automatic sprinkler controllers or garage door openers (these take surprisingly more energy than you think). Work with your baseline number until you have a rough understanding of everything that goes into it.

# Auditing Your House

In this section, I explain how to systematically go through your house and rectify the most costly energy problems. If you do this audit yourself, you can understand what makes your house tick. Make sure that you put on your grungiest clothes, because you're going to be climbing around in your base-ment and attic. (But please be careful.)

The point is to do something about your big inefficiencies, not to study all the little ones. In economics, the idea behind this strategy is called *diminishing marginal utility*. It says that the first dollar you spend on improving your effi-ciency will have a big impact. The next dollar you spend, not so big an impact. After you get out to a certain point, you no longer get a dollar's return for your dollar investment, which is when you should stop — unless you're working with carbon footprints, of course. Then it's up to you.

You can get professional help if you need to (see the upcoming "Getting Professional Audits" section). Plenty of resources are available. But at the very least, look over the shoulder of the professional if you decide to hire one. Ask questions, and find out what's going on. If he or she says some-thing's amiss, look for yourself.

## Plugging leaks

You can save from 5 to 30 percent off your heating and air-conditioning bill simply by plugging up air leaks, as I describe in the following sections.

Check for gas leaks as well. Gas system leaks are dangerous and costly. You should be able to smell them. If you do smell a leak, call a qualified service technician from your utility company as soon as possible; gas leaks can be very dangerous.

## *Pressure test*

You may already know where the drafts are. Find out why. Is the air entering through an unsealed door? A window? A vent? Is it coming from a heater vent? A pressure test can help you pinpoint those leaks so you can trace them down and seal them off. Try to perform this test when it's cold outside and warm inside.

1. **Seal off your house and turn off heating and cooling sources.**

   Completely extinguish any fireplace fires. Close the fireplace damper as much as possible. Turn off your HVAC system. Turn off any furnaces. If you have a gas water heater, turn that off, too. Close all the windows and doors in your house. Make sure to close any skylights or vents.

2. **Turn on all the exhaust fans in your house (normally located in kitchens, bathrooms, and laundry rooms).**

   If you don't have any exhaust fans, aim a portable fan out a single open window and turn it on.

   Quickly make sure that your fireplace is okay. If it's leaky (for example, air is coming down the chimney and out into the house), you'll be drawing in some stink. If so, turn the fans off and inspect your fireplace to find out why it's so leaky. If you can, fix it because it's inefficient (when you're not using the fireplace, heat will be escaping up the chimney). If you can't fix it, forge ahead if the smell isn't too bad. If the smell is really bad, may want to call a fireplace specialist. Old houses often have very leaky fireplaces because people simply didn't care about energy efficiency back in the old days. Maybe it's time to install a modern, well-designed gas stove.

3. **Search for leaks around the house.**

   With the fans on, your house is depressurized, so any leaks are readily apparent. Go around the house with a bowl of water, dip your hand in, and move your wet hand around windows, electrical outlets, switches, doors, molding interfaces, attic hatches, basement hatches, and so on. You should be able to feel a leak, especially if it's cold outside.

   Another way to do this step is with a stick of incense; when the smoke fluctuates, you've found a leak. Or use a candle; when the flame flickers, you've found a leak.

4. **Get a ladder or a chair and check for leaks in overhead lights.**

   Such leaks are very common, but unfortunately they're a little more difficult to fix if you don't have good attic access.

5. **Turn off the fans and then fix the leaks as appropriate.**

Caulking works wonders, but it can look bad. You may want to use the clear kind. Or perhaps you can fix the leaks from the outside, where caulking is more visually acceptable. Drafts start somewhere and end somewhere, so you're best off fixing both sides of the problem. You may be able to seal off some areas with duct tape.

Buy a caulking gun that has a pressure relief gasket, or else the caulk will keep coming out when you're done squeezing the trigger, making a big mess (and possibly leading to PG-13 language).

You can also buy aerosol cans of expandable insulation that work really well, but beware; this stuff is nasty, and it gets all over everything. I have yet to find anything that can clean it off, but it sure works well. Your best bet is to go in your garage first and practice spraying it on a newspaper you can ball up and throw away.

If you have a leak in an outlet or switch, turn the electricity off in the entire house before you go in there and fix it. Flip the main circuit breaker off. You may have to reset your clocks, but you'll be alive to do it — so don't complain!

6. **You may need to climb into the attic to fix overhead light fixtures.**

Step only on the joists; if you step on the sheetrock, your foot may go down into the room below, which would be very startling for all concerned. Or you could go down into the room below, which would be even more startling!

7. **After you've finished fixing leaks, turn the exhaust fans back on and repeat the wet hand (or incense or candle) routine from Step 2.**

Your house will be tighter now, and any remaining leaks will be even more obvious. Repeat the fixing stage.

8. **Repeat the whole process until you're satisfied.**

### Door seals

Check the seals around your exterior doors. Applying foam weather stripping is easy; it comes in self-stick tapes of various sizes. Measure how much you need, pull off a chunk of the existing seal (if there is some), and take it to the hardware store as an example of the specific type of material you need. There are a lot of different forms and thicknesses; don't go to the hardware store expecting to guess right. If there is no seal, measure the surface widths that you will applying the material to, and get the hardware store clerk to help you identify the right material. The bottom line is that some weatherproofing, even if it's not optimum, is much better than none. Your worst problem may be appearance. Use a sharp box knife to cut off any excess material.

The DOE's Consumer Guide (www.eere.energy.gov/consumer/your_home/insulation_airsealing/index.cfm/mytopic=11280) may be helpful for deciding which types of insulation work best in your climate and physical conditions.

Check along the bottoms of the doors as well. Sealing thresholds takes a little more effort, but your hardware store likely has plenty of insulating solutions that you should be able to implement yourself with no more than a screwdriver. Check out the store's stock and how the different items are used. For example, vinyl door sweeps are very effective for the bottom of a doorway, while foam insulation works well around the periphery.

### Window seals

Stationary windows should be well caulked. Get the good stuff, the kind that lasts for 50 years. (You don't need it to last for 50 years, but it'll work better for 10 years if it's rated for 50.) Sliding windows take more ingenuity, but your hardware store clerk can help you.

### Heating, ventilation, and air-conditioning inspection

Most ducts are for sending heated or cooled air into the house; one large one is the return. They all need to be tightly sealed. Leaks in the ductwork are worse than air leaks in your house because the ducts are pressurized, which magnifies the amount of air escaping through cracks and openings. So make sure that you access your HVAC system and visually inspect the duct system snaking around your house. If the ducts aren't well insulated, you can get kits at hardware stores. And for additional advice, check out *Energy Efficient Homes For Dummies* (Wiley) by yours truly or visit The U.S. Department of Energy's guide to energy efficiency at www.eere.energy.gov/consumer.

When working with dusty ducts, wear a dust mask; get them at your hardware store. (By now, you should know where that is.)

You can find many problems with just a quick glance. It's amazing how many people are heating their basements without even knowing it. I've been in a lot of houses where the insulation is just shredded off the ducts. I've also been in a lot of houses where junctions have broken, leaving a big opening. Rats and mice like to chew through, and they leave little openings that are hard to find (unless you're a small rodent).

Here are some things to remember when you inspect and fix your ductwork:

- ✔ You can turn the HVAC on so that the ducts are pressurized and find leaks with a wet hand. (See the earlier "Pressure test" section for info on how this works.)

- ✔ Duct tape works wonders, but it doesn't stick to dusty surfaces very well. If you have a dust problem, wrap the duct tape around and around and just cover up the dust. It's easier than trying to clean it off.

- ✔ Close the ducts leading to rooms where you don't need the HVAC system to operate. Most of the time this is very easy: Simply slide the tab that opens and closes the vent. If it's more complicated, ask how to do the job at the hardware store; it should have the materials you need. If you don't care about appearances, duct tape will do the trick.

### Outdoors now

Go outside and visually inspect for leaks. Use your imagination. And remember: When in doubt, squirt caulk! What's the worst that can happen? Here are some places to pay special attention to:

- ✔ **Look at faucets, pipes, electrical wiring, and electric outlets.** Cracks often form around the junctions where the pipes fit through foundations and siding; fix these with caulk. Even if you're only sealing off your basement, which you're not heating or cooling, you'll be better off inside the upper part of the house .

- ✔ **Check all interfaces between two different building materials.** Bricks to foundation; interior corners with molding strips; where siding and foundations meet; roofs to siding; and so on. Plug all holes and voids with caulk — the good stuff.

If icicles are clustering around a particular location at your house, you have a leak somewhere above that's melting snow. The water drips down and then refreezes into icicles. These leaks are usually pretty good sized and easy to locate and fix.

- ✔ **Look for cracks in mortar, foundations, siding, and so on.** Seal these with appropriate materials.

- ✔ **Check for cracks and voids around exterior doors and windows.** These gaps may not result in air leaks inside the house, but while you're in your grungy clothes and the proper mood, you may as well seal water leaks to prevent damage that could cost money and turn into air leaks.

- ✔ **Check storm windows for seal integrity.** The interior window may be well sealed, but the storm window will work better if it's also sealed.

## Checking insulation

Check insulation wherever you can. If you find it wholly inadequate, you have a big decision on your hands. Putting in new insulation is expensive, and the payback time is going to be way out there. On the other hand, you're wasting money and energy if you don't do it. Here are some areas you can focus on:

- ✔ Voids in the insulation are easy to deal with. They usually occur around light fixtures and where somebody has been working, pushing it aside. Fix these voids by either replacing the insulation, or filling them with expandable foam insulation (the real messy stuff).

- ✔ The most important room is your family room, so if you only have enough money to insulate one room, that's it.

- ✔ If you have a basement, check to see whether the ceiling is insulated (that would be under the floor in the room above). If not, putting insulation in is a relatively simple job and very cost effective. One of the most cost effective insulation projects is to insulate beneath the family room.

✔ Having the insulation in the attic thickened is easiest. You can do this yourself, although working with insulation is a hassle (make sure that you use a dust mask at all times). Simply put more insulation over the existing stuff. You can find companies in the phone book to come in and spray powdered insulation, which works nicely.

✔ Hot water pipes should be well insulated. A number of easy options that work well are inexpensive and widely available. The best bet is a length (usually four feet) of foam cylinder with a lengthwise cut that allows you to simply push the piece over a pipe. Another option that works a little better is fiberglass insulation tape, but it requires you to unroll the tape around and around the pipe, which can be very difficult if the pipe is attached close to a wall, or under a floor. Ask at your hardware store.

When the insulation in your attic is substandard, it's probably also substandard in the walls, and that problem isn't so easy to remedy. The best bet may be to simply put new siding on the outside of your house. You can get good-looking siding with great insulation properties, and when you spend money like this, you can accomplish a visual remodeling as well.

## Minding the details

The following list covers some details that can make a big impact, depending on your home.

✔ **Increase window insulation.** Single-pane windows are poor insulators. Changing to double pane is best but expensive. You can put up heat-sealing cloth barriers in the summer or storm windows in the winter. Check out Chapter 9. This step can make a very big difference, and you don't have to spend an arm and a leg. You may consider doing it in one room only, such as the family room or your master bedroom.

✔ **Let attics breathe.** Attics need to breathe properly. They're usually built with vent systems either right through the roof or up in the walls below the eaves. If the vents get completely clogged with dust, your attic gets much hotter in the summer. Clean out all vents (the easiest way is to use one of those extension poles commonly sold for spider webs). If you don't have any vents, consider putting some in. Ask at your hardware store to find out how best to do it in your area.

✔ **Change HVAC filters.** Clogged filters force the machinery to work harder, which takes more energy to do the same job. Changing filters every month or so is unquestionably cheaper than paying the extra power bill. You can get really fancy, expensive filters, but don't bother — they're not worth it. Buy a whole box of cheap ones (much cheaper by the dozen) and change them frequently. With some types of filters, you can simply vacuum the dust off and they'll work like new. Don't do this more than two or three times, however.

Filters in your range hood can get gummed up with grease, preventing your fan from pushing any air. The fan can end up spinning the power meter like mad, with minor effect, so make sure that you change range hood filters as well. Or you can clean the existing filter with a kitchen degreaser. Let it soak overnight; then rinse it off and put it back where it belongs.

✔ **Replace or service inefficient HVAC systems.** If your HVAC systems are old, they're undoubtedly inefficient. Your best bet is to call an HVAC service company and have someone come out to analyze your equipment. The serviceperson can tell you how much better new equipment will perform, although you should keep in mind that his or her motive is to sell new equipment. Do your research, and let the buyer beware. Get at least three quotes.

If your systems aren't old, having them serviced still may pay (with the realization that this may be asking for a scam). My experience with HVAC service companies is that many have a set price, regardless of what they do. A broken heat pump costs $267. A new fan costs $267. Tightening some screws? You guessed it. Before anyone comes out, ask him how he charges and what you're going to get for your money. Does he warrant his work? If he fudges, call around. You can find somebody who's on the up and up.

✔ **Lower your wattage, and turning off the lights.** Each light bulb is clearly marked with a wattage. A 60-watt bulb left on for an hour consumes 0.06 kWh. Ten 60-watt bulbs in recessed lighting in your ceiling turned on for 4 hours consumes 2.4 kWh. At a rate of 15 cents per kWh, this costs 36 cents a day. For a month, the total comes to $10.80, or $130 per year. Add to that the cost of new bulbs, and it may be costing you more than $150 a year to leave those overhead lights on every night. Calculations like this can be surprising.

Do you have an outside light that burns all night? If it's 600 watts, it's costing you:

$$0.600 \text{ kWh} \times 10 \text{ hrs/day} \times 30 \text{ days/month} \times 0.15 \text{ \$/kWh} = \$9\text{/month}$$

Don't cut yourself any slack if you use dimmers. They're very inefficient.

Fluorescents use much less power to put out the same light intensity. They cost more (but last up to ten times longer, so not really), but compare this expense to the gain you get not only with reduced power bills but also a reduced carbon footprint (see Chapter 1 for details on carbon emissions). Okay, so fluorescents look different; they have a bluish light that flickers. You can't put them on a dimmer. Sometimes they're just not right. But at least give them a try.

## Analyzing your major appliances

Major appliances consume a lot of power. To find out how much an appliance costs per month to run, first estimate how much time it's on per day. Then use this formula:

Wattage $\div$ 1,000 $\times$ hrs/day $\times$ \$/kWh $\div$ 30days/month = total cost per month

A clothes dryer uses 5,570 watts. If you dry clothes for six hours a week, that's 6 hours $\div$ 7 days = 0.86 hours per day. Here's what a month's worth of use would cost you:

5,570 $\div$ 1,000 $\times$ 0.86 $\times$ 0.15 $\times$ 30 = \$21.56 per month

If you iron clothes each morning for 15 minutes, it's costing you

1,200 $\div$ 1,000 $\times$ 0.25 $\times$ 0.15 $\times$ 30 = \$1.35 per month

Add to that the cost of your washing clothes for six hours a week:

900 $\div$ 1,000 $\times$ 0.86 $\times$ 0.15 $\times$ 30 = \$3.48 per month

Your total for clothes is \$26.40 per month. If you put up a clothesline, you save \$21.56 per month in cash and about 1,748 pounds of carbon dioxide emissions per year. Plus, your clothes smell better. You may also want to consider that all that dryer heat eventually goes out the vent pipe to the great outdoors, which seems like such a waste.

# Getting Professional Audits

You can call your power company to do an audit for you. Some companies come to your house and look at your situation in detail — these are the best kind. Other utilities offer an Internet analysis or a mail-order analysis, which is probably only of modest interest because it doesn't take your specifics into account. Of course, if it's a mail-in type deal, you may as well just struggle through it yourself — by the time you're done compiling enough information to make the mail-in audit worthwhile, you've basically done the job yourself. Same thing with Internet audits.

If you get a professional audit done, take some of their suggestions with a grain of salt. The biggest inefficiencies are easy to detect, especially if you've never even tried before. I can tell you from experience doing audits that most houses have glaring problems that can usually be fixed for less than a few hundred dollars.

If you do decide to hire an auditor, look here:

- ✔ National Association of Energy Service Companies (NAESCO) 202-822-0950 or www.naesco.org
- ✔ The phone book under energy conservation services and products
- ✔ Phone book under utilities, electric service or utility providers, or gas utility companies
- ✔ The Home Energy Saver (http://hes.lbl.gov), sponsored by the U.S. Department of Energy, which offers a free home energy audit as well as carbon footprint estimates

Here's what to do prior to a professional audit:

- ✔ Assemble all electric and utility bills for the past two or three years.
- ✔ Make a list of occupant habits, especially noting whether anyone is home during the day. Are there kids? Are clothes washed and dried frequently? Are long showers taken? Big baths?
- ✔ Note the thermostat setting in your house.
- ✔ Make a list of questions you may have.

    • If you want to make a change or improvement in your household, ask how it would fit in with the results of your audit. Then ask for some advice. Auditors have tons of experience, and they're usually very proud to expound.

    • Ask about financing programs the company has for improvements the auditor suggests.

    • Ask about guarantees for their work. If they tell you that you can achieve a reduction, how accurate are they warranting that claim?

# Chapter 3

# Making Your Home Energy Efficient

*B*efore installing solar power, reducing your energy consumption as much as possible makes sense. For every dollar you spend on energy conservation, you'll save much more on the cost of your solar system. For example, if you can save 1 kWh per day on your power bill (5 percent of the typical North American household's energy use), your solar energy system will cost around $7,000 less. The number depends on the level of subsidies and tax breaks you can take advantage of (see Chapter 20), but the impact is obvious. The term that applies to conservation is "negawatts," which derives from the word "negative." When you conserve, you are negating your energy use, and that's easily the best way to save money and help the environment.

In this chapter, I show you how to make your home energy efficient. Some ways are cheap and easy; some are expensive and involved. Go ahead and take your pick. For additional advice, visit the U.S. Department of Energy's guide to energy efficiency at www.eere.energy.gov/consumer. And for a more detailed account of making your home energy efficient, buy or borrow a copy of my book *Energy Efficient Homes For Dummies* (Wiley). In fact, buy ten copies and hand them out to family and friends; they'll love you for it.

It's been my experience that through conscientious change of habits, the average household can reduce its energy consumption 20 percent. And through investments in energy-saving equipment, another 15 percent may be possible. That's a third off your monthly bills, and it'll reduce the cost for your solar investments because you can install less capacity! (Need I mention pollution? Didn't think so.)

Get a copy of the latest Real Goods catalog (www.realgoods.com) for all kinds of toys and gadgets so you can get a feel for the kind of consumer equipment that's available.

# Conserving Energy without Reducing Quality of Life

Conserving energy reduces greenhouse gases, relieves the strain on natural resources, lowers energy costs, and eases the manmade assault on Mother Nature. There seems to be an inherent assumption that energy conservation also entails a reduction in quality of lifestyle, but Europeans use far less energy than North Americans, and arguing that their lifestyles are any worse than ours would be difficult. In Europe, energy efficiency has been ingrained for a long time to the point where it's a self-perpetuating logic of its own.

On a residential street, most of the houses are similar. The people living inside don't look the same, but culturally, they're similar. They're in the same neighborhood, after all. But one house may be paying $400 a month in power bills while the other is only paying $50. You can see an equivalent difference in their carbon footprints. (For more on carbon footprints, see Chapter 1.) Yet by all external accounts, both houses enjoy the same quality of life.

The house with the lower power bills has solar equipment installed on its roof. And it probably has deciduous trees on the southern front. It looks better for it, especially in the autumn. It has solar light tubes in the kitchen so that lights aren't on very often. The occupants open and close windows to optimize ventilation. They use space heaters in winter and turn their thermostats down, but they wear sweaters so they're just as warm.

Here are some more things you may notice about the efficient house:

- ✔ It's darker at night.
- ✔ The garage door is closed all the time.
- ✔ Blinds and awnings are opened and closed a lot, depending on conditions.
- ✔ Solar tubes and skylights are visible on the roof.
- ✔ A clothesline bisects the backyard.
- ✔ The noise pollution from an HVAC (heating, ventilation, and air-conditioning system) is much less because it's used much less.
- ✔ You don't hear a TV on all the time.

You get the picture. Both houses have the same quality of life, but there's a very big difference.

# *Changing Habits and Equipment Is Hard to Do — Or Is It?*

This section discusses some of the easiest ways to save power. You don't even need to make a cash investment. All you have to do is change some habits and become more aware of your consumption. In Chapter 2, I describe ways of auditing your consumption. Simply by being more aware of how you use energy, you'll find all kinds of ways to conserve.

## *Lighting*

Most households have a good number of lights on at the same time (some even during the daylight hours!). Basic logic suggests that you don't need more than one light per person turned on at any give time, but let's be practical — lighting affects ambience quite a bit. It can make a room seem much more inviting, softening the hard edges and implying security and well being. Lighting can magnify the best features of your house, both inside and out, and it's not reasonable to reduce lighting to the strict task of functionality.

### *Lighting (and darkening) the interior*

You can probably do better with your lights. Do you turn them on by habit or necessity? Sure, the house looks nice and warm with a bunch of incandescents burning away, but you're an emancipated energy slave now, and it's time to revert to darkness. Let there be dark!

Take a look at some tips:

- Here's some bottom-line logic you can't argue with: One light bulb per person in on the house at any given time.

- If you want a dimmer, just use lower wattage bulbs. Dimmers don't save energy; they just consume it differently. Experiment with lower wattage bulbs; if you don't like the result, go back. But you'll probably find that lower wattages work just fine. In fact, lower wattage bulbs tend to have a warmer tone, so odds are you'll like them. Since they run cooler, they last longer as well. You save on your electric bill, and you also save by not having to change your light bulbs as often.

Incandescents are inefficient because they put out a lot of heat along with their light (a typical incandescent is only 10 percent efficient; in other words, a 100-watt bulb is putting out 90 watts of heat and only 10 watts of light; that's the reason so many utilities are encouraging and subsidizing the use of fluorescents). Using incandescents makes even less sense in the summer, when you're trying to cool your house. For every 100 watts of incandescent lighting, your air conditioner needs to run an extra 80 or 90 watts to offset the heat. Using fluorescent bulbs in the summertime can make a big difference.

The reason many people don't use fluorescents is habit. So maybe you don't like the bluish, flickering light. It looks like an office, and you get enough of that at work. But newer versions are much better. The spectrum is friendlier.

Fluorescent bulbs contain mercury, so when they finally, *finally* burn out, recycle them or dispose of them at a hazardous waste facility.

### Lighting the great outdoors

You don't need to limit your energy conservation habits to the inside of your house. For example, many people like the sense of enhanced security with bright outdoor flood lighting. A more sensible approach is to use a motion detector to turn the light on for only five minutes at a time. You still achieve your security — probably even more so because an intruder is more likely to flee if he's suddenly illuminated. He (okay, *she,* but that's not as likely) may even be more inclined to intrude if the light is on all the time because he or she will be able to see that no threats are around, such as sleeping dogs, people in a corner on a lawn chair with automatic weapons, and so on.

A number of inexpensive products on the market feature the following:

- Motion detectors and timers to limit the on-time

- LEDs (Light Emitting Diodes) for bright spotlights at very low energy consumptions

- Solar power systems that don't need any hardwiring (You can get the same performance as a hard-wired overhead light. The only drawback is that you need to mount the solar collectors where they'll get some sunlight, but as I explain in Chapter 5, this is easy to do.)

## Appliances and electronics

Appliances gobble up a lot of energy. That's one reason why the Department of Energy, in conjunction with the Environmental Protection Agency (EPA), has devised an energy efficiency rating system named ENERGY STAR that tells you how much energy you can expect an appliance to consume in a year. Figure 3-1 shows you a sample appliance label that uses the ENERGY STAR rating system. (Visit www.energystar.gov for more information.)

Beneath where it says "This Model Uses" is the estimated energy consumption for the particular model. Below that amount is the minimum energy consumption for models of the same type, as well as maximum energy consumption for models of the same type. You can tell how good (or bad) your appliance is with respect to the industry standards. Of more interest is the annual cost, which is estimated using national energy cost averages. If your costs are higher than the national average, expect your appliance to cost more, and vice versa.

You can use these numbers to calculate how much money you'll save by throwing out your old appliance (donate it to a charity and write it off, or sell it in the newspaper) and replacing it with a nice new one. Don't forget to calculate carbon footprints (see Chapter 1). New appliances are often quieter and better looking as well.

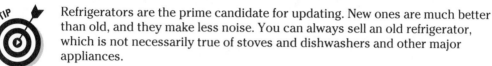 Refrigerators are the prime candidate for updating. New ones are much better than old, and they make less noise. You can always sell an old refrigerator, which is not necessarily true of stoves and dishwashers and other major appliances.

Buying new appliances isn't the only way to rack up savings. The following sections discuss ways you can reduce energy usage with your current appliances and electronics.

### Cooking appliances

Limit heat from cooking. You can turn your oven off before food is finished cooking, especially in the summer. Don't preheat your oven, like it tells you to do in the recipe. It doesn't really make any difference in the quality of the final product; you just have to pay more attention to when the food is done. And in the summer, whatever heat you generate in your oven will eventually bleed into your house, so that exact amount of heat will have to be offset by your air conditioner. You also get this added heat when you cook on the stovetop.

 Use the barbecue in the summer to cook all your food, including vegetables. Gas barbecues are much better for the environment than charcoal, and because they're much more convenient, you'll be more inclined to use them. And contrary to what a lot of barbecue aficionados claim, gas barbecues don't smell bad. Besides, most people who use charcoals squirt a bunch of nasty lighter fluid over the coals, which is terrible for the environment and in some communities, it's banned altogether. If you insist on charcoal, use an electric starter instead of lighter fluid.

Microwaves are the most efficient way to heat foods, but sometimes they're not practical because of the astringent textures that result. Toaster ovens provide a happy median, and I recommend sitting one next to your microwave. Get one with a timer, or at some point, you'll forget what you're doing, burn something, set off the smoke alarms, and wake up the dog.

Or you can use a solar oven, which effectively captures solar radiation into a confined, insulated space to heat just like a conventional oven but with no energy cost and no pollution (see Chapter 9).

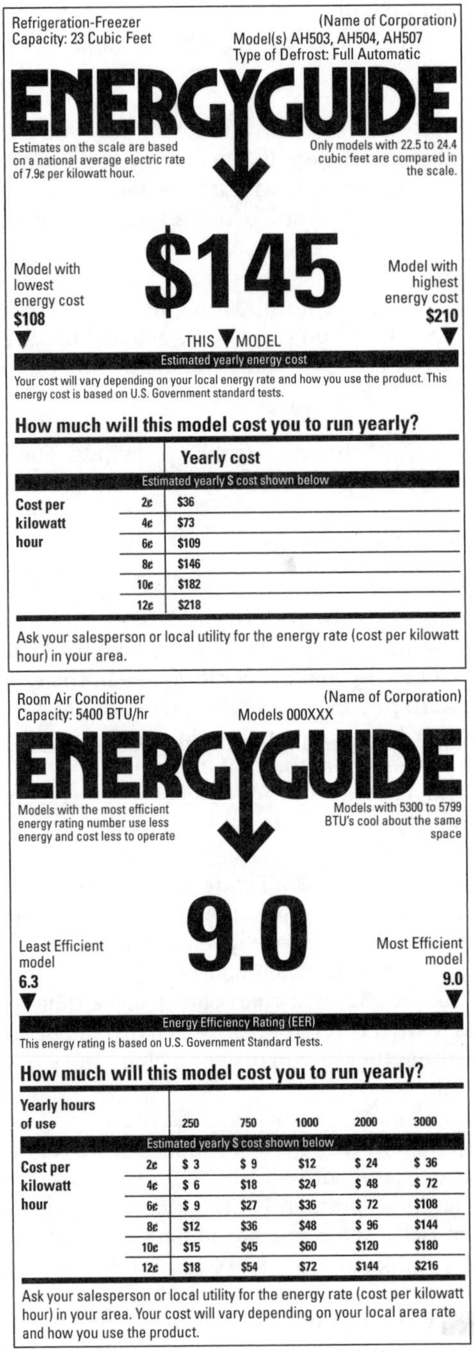

Refrigeration-Freezer
Capacity: 23 Cubic Feet

(Name of Corporation)
Model(s) AH503, AH504, AH507
Type of Defrost: Full Automatic

# ENERGYGUIDE

Estimates on the scale are based on a national average electric rate of 7.9¢ per kilowatt hour.

Only models with 22.5 to 24.4 cubic feet are compared in the scale.

Model with lowest energy cost
**$108**
▼

## $145

THIS ▼ MODEL

Model with highest energy cost
**$210**
▼

Estimated yearly energy cost

Your cost will vary depending on your local energy rate and how you use the product. This energy cost is based on U.S. Government standard tests.

## How much will this model cost you to run yearly?

|  | | Yearly cost |
|---|---|---|
|  | | Estimated yearly $ cost shown below |
| Cost per | 2¢ | $36 |
| kilowatt | 4¢ | $73 |
| hour | 6¢ | $109 |
|  | 8¢ | $146 |
|  | 10¢ | $182 |
|  | 12¢ | $218 |

Ask your salesperson or local utility for the energy rate (cost per kilowatt hour) in your area.

---

Room Air Conditioner
Capacity: 5400 BTU/hr

(Name of Corporation)
Models 000XXX

# ENERGYGUIDE

Models with the most efficient energy rating number use less energy and cost less to operate

Models with 5300 to 5799 BTU's cool about the same space

Least Efficient model
**6.3**
▼

## 9.0

Most Efficient model
**9.0**
▼

Energy Efficiency Rating (EER)

This energy rating is based on U.S. Government Standard Tests.

## How much will this model cost you to run yearly?

| Yearly hours of use |  | 250 | 750 | 1000 | 2000 | 3000 |
|---|---|---|---|---|---|---|
|  | | Estimated yearly $ cost shown below | | | | |
| Cost per | 2¢ | $ 3 | $ 9 | $12 | $ 24 | $ 36 |
| kilowatt | 4¢ | $ 6 | $18 | $24 | $ 48 | $ 72 |
| hour | 6¢ | $ 9 | $27 | $36 | $ 72 | $108 |
|  | 8¢ | $12 | $36 | $48 | $ 96 | $144 |
|  | 10¢ | $15 | $45 | $60 | $120 | $180 |
|  | 12¢ | $18 | $54 | $72 | $144 | $216 |

Ask your salesperson or local utility for the energy rate (cost per kilowatt hour) in your area. Your cost will vary depending on your local area rate and how you use the product.

**Figure 3-1:**
ENERGY
STAR label.

## *Refrigerators*

Here's how to keep both the temperature and your energy use down:

- ✔ **Use your ice dispenser, if you have one.** You won't have to open your door and let in the warm air, which then needs to be recooled, costing extra energy.

- ✔ **Turn the temperatures up on your existing freezer/refrigerator.** Experiment to see how it works. It makes a big difference in cost but not much in terms of food quality.

- ✔ **Clean your coils.** At the bottom of your refrigerator are cooling coils that need to be vacuumed at least twice a year. When crud builds up on the coils, your compressor is much less efficient, which means your refrigerator has to run longer. Does your refrigerator run all the time? Dirty coils are probably why. You can get special dusters from your hardware store. After cleaning the coils, not only do you save on energy, but your house is also quieter. It's amazing how many people never, ever clean their coils (the same set who never change their HVAC filters).

- ✔ **Get rid of that old freezer in your garage.** Why is it there again? Because you can get a price break when you buy an entire pig? Is it worth filling your energy pig with a pig? Consider carbon footprint.

## *Electronics*

Look around your house and you'll find all sorts of electronic equipment that's drawing power even when it's turned off. LEDs are illuminating, fans are running. You don't think this adds up? Turn off everything in your house and go out and look at your electric meter. For most of you, it's still spinning. The best way to control phantom loads is to use power strips (the kind that are common with computer systems). You can get them at any variety store for less than ten bucks; make sure to get one with grounding prongs and quality switches. When you're not using the equipment, simply turn off the power strip. This works especially well for entertainment centers. Big TVs and stereos draw a lot of current even when they're shut off. Your cable box also draws a lot of current even when turned off. In this day and age, "off" does not really mean "off."

Many people flip their television on as soon as they get home. It makes an empty house less lonely and enlivens the environment. If you need that external stimuli, try a radio instead. It consumes much less power. Or maybe you might talk to your kids. If they're not more interesting than TV, it's probably your fault, not theirs.

# Testing, one two three

Here's a fun little project. See what it takes to get your meter to stop dead. You may not even be able to do it. Try shutting off your entire house by flipping down the main circuit breaker in your fuse box. Is your meter still moving? Call the power company because you may have a current leak through the ground. It happens.

If you shut off your entire house and your meter stops moving, but then when you turn the power back on you still have movement when everything's turned off, start unplugging things.

First unplug the little gizmos with LEDs, such as your cable box. Telephones. Stereos and clock radios. Unplug your refrigerators.

Check out how much energy each item uses. A lot of meters have outdoor outlets located nearby. If not, run an extension cord over there. Experiment with some of your favorite appliances and gadgets. For example, plug in your hair dryer and watch your meter go berserk. Try a space heater; they gobble juice up about as fast as Speedy Gonzalez.

## Vacuuming and laundry

Here are some tips on efficiently using the appliances that keep your carpet and clothes clean and dry:

- **Get an upright vacuum cleaner.** Central vacuums waste tons of energy, and they move far more air than an exhaust fan. (The good news is they're hard to leave on too long because they make so much noise.)

- **Forget the dryer and use a clothesline instead.** You can get nifty retraction mechanisms to make the clothesline disappear when not in use. You don't even have to air your dirty — uh, clean — laundry outside. Just hang the clothesline in your garage and run a fan near an open window while you're at work (unless your garage is full of stinky old cars or incontinent cats!). You can also hang certain items of clothing over the shower rod in your bathroom. Open the window, or drying may take forever.

- **Always keep your dryer's lint trap cleaned out.** Every time you run the dryer, check it. Yes, every time; it makes a big difference.

- **Clean clogged dryer vents.** A snaking, crimped dryer vent hose is as bad as a clogged filter — maybe even worse. The vent hose may be very difficult to check out, but it's worth it — clogged vents are also a fire hazard. I've seen dryer hoses filled with lint — no air movement at all. No wonder the dryer was on for three hours at a time, and then the clothes were still damp and hot.

# *Hot water*

Whether you're heating water in the sun (see Chapter 10), using solar systems to supplement your domestic water heater (Chapter 12), or just trying to cut down on how much energy your water heater uses, you can easily reduce your hot-water needs — without kicking anyone out of the house.

### *Practicing some slick hot-water conservation habits*

Hot water consumes an average of 18 percent of a typical household's energy needs. You can make a big difference in your power bill if you can just manage to change a few habits. Most of these suggestions are easy to implement, and you'll quickly find that your quality of life has been impacted only marginally.

- ✔ **Use cold tap water.** Most of the sink taps nowadays have a single lever for controlling both hot and cold water. You push the lever to one side to get hot and vice versa. Rarely do people actually want hot water, but inevitably they push the lever to the middle when they draw water for any use at all. Even though you're not getting any hot water from the tap because it takes a while to get there, you're still drawing hot water from the tank and wasting it. From now on, when you draw tap water, push the lever completely to the cold side.

  On that subject, drawing hot water from the kitchen tap for use in cooking is wasteful. When you draw hot water, you need to wait a few minutes for it to get there. You're filling up all the intervening pipes with hot water in the process. Then you fill a cup, or a bowl with the hot water. The fact is, the amount of water in your pipes can be over ten or twenty times the amount you actually use. The heat dissipates and is wasted.

  Heat cold water in the microwave oven. The stove is better than the tap, but not as efficient as a microwave.

- ✔ **Run a full dishwasher.** Washing a 12-piece dinner set of dishes by hand takes around 2.5 kWh of heated water. Washing the same dishes in a machine takes only around 1.5 kWh, and the dishes are much cleaner. You don't need to rinse the dishes with hot water prior to putting them into the dishwasher. Most people prewash their dishes more than they need to.

- ✔ **Take showers instead of baths.** A five-minute shower takes a third of the water of a bath. If you want to spend half an hour soaking, a bath is better, but most people simply want to clean off.

✔ **Lower the water temperature at the source.** Set the temperature of your domestic water heater to 120°F instead of the scalding 160°F it's probably at right now. Here's a table of the costs of heating 64 gallons of water per day with electric power when you set the thermostat to different temperatures:

| Temperature | Yearly Cost (at 12 cents per kWh) |
| --- | --- |
| 160°F | $675 |
| 150°F | $604 |
| 140°F | $537 |
| 130°F | $460 |
| 120°F | $405 |

That's a typical savings of $270 per year. Will you even be able to tell? Probably not. So why is it set so high? It's the old school of thought, from back in the days when energy costs were inconsequential (not to mention concern about the environment). These were the same days when people smoked a lot of cigarettes because they were good for your health. I wonder if our children are going to look back and think that we were as ignorant as we perceive our forebears to be.

### Fixing up your shower, pipes, and water heater

You can also save energy just by making a few household repairs or improvements:

✔ **Repair leaky faucets, especially hot water.** The leak may seem tiny, but the cost adds up fast.

✔ **Put a flow constrictor into your shower head.** And take a look at how much water your current shower head is wasting. Is water going all over the place? What's the point in aiming water at the walls? You also don't need to turn the hot water up so that the windows steam. You can take a longer, cooler shower and save energy. (Of all the advice I give out about energy conservation, this is ignored more than any other. Many people simply can't live without their long, hot showers. For these, a solar water heater is in order.)

Get a gallon container and turn your shower on. Measure how long it takes to the fill the container. Your flow rate is 60 divided by the measured number of seconds. If it takes 15 seconds, your flow rate is 4 gallons per minute. A typical rate is 2.5 gallons per minute. A good energy conservation rate is around 1.75 gallons per minute. You also may want to measure the capacity of your pipes. Use the same container to measure how much water flows through your shower head before it starts to get hot. When you're finished with your shower, that much hot water is now sitting in the pipes, and the heat is wasted into the environment.

- ✔ **Drain your domestic water heater twice a year through the valve at the bottom to remove water heater crud.** Drain about two quarts into a bowl or container. You'll get rid of the sediment that settles on the bottom that makes for less efficient heating. In most cases, the crud is very obvious; it's a muddy, grey texture. And yes, this stuff gets into your water system, which means that it gets into your hair and your food.

### Using on-demand water heaters and heat recovery systems

An on-demand water heater doesn't have a large reservoir of hot water. Instead, it uses a small heating chamber with super high-speed capacity. As water is drawn through the chamber, it heats it up to the set temperature within seconds, so you suffer no heat loss as the water sits there in a storage tank. This heater makes more sense in houses where you don't use hot water for extended periods of time, such as in a vacation home that's empty for entire seasons.

Look in the phone book for suppliers in your area. These systems are more expensive than conventional, but if you use hot water only occasionally, they'll likely pay off.

A cheap alternative is to simply use a hot water tank blanket, which is a layer of insulation that you can buy for around $20 at most hardware stores. Write down the model of your hot water tank, and a clerk at the store can identify the size of the blanket that works best.

# Heating, ventilation, and air conditioning

Heating, ventilation, and air conditioning (HVAC) comprise, on average, around 25 percent of a typical household's energy costs. Your proportion may be much greater, depending on your climate. And most homes are inefficient in terms of insulation properties, so there is a lot of wasted energy. By changing a few habits and making some inexpensive modifications, you can save a lot of money.

### Keeping your cool as you change temp control habits

By simply altering the way you run your equipment, you can save money on energy costs. Habits die hard, it's true. But you'll find that these changes won't be nearly as hard as, say, eating much less, or exercising much more.

- ✔ **Make sure your registers flow freely.** How much junk can fall through and create air dams is amazing. Children's toys are prime candidates. You can hurt your heat pump if all the registers in your house are shut off or unnecessarily blocked. What's the point?

- **Close off unused rooms.** You can do so by either closing the registers (or turning off the valve if you have a radiator) or by closing the door to that room.

- **Cut your use of exhaust fans.** When you run bathroom or kitchen exhaust fans while your HVAC is working, you're pulling the conditioned air from your home and forcing it outside. This air needs to be replenished somehow. You draw in hot or cold air through leaks in your home's envelope, and the HVAC system needs to work a lot harder than necessary (which means you're wasting money).

  Do you really need to use that exhaust fan? If so, crack open a nearby window to minimize the wasted energy. After you're done using the fan, close the window again.

  Sometimes in bathrooms the exhaust fan is connected to the light, and whenever you turn the light on the fan is also on. Face it: There's only one good reason to use an exhaust fan in a bathroom. Yes, you know when. The rest of the time it's a waste (no pun intended). You can also put an exhaust fan in with its own switch, maybe even remotely controlled. Ask at your friendly hardware store for the different options.

- **Use dehumidifiers in hot climates.** Dehumidifiers make a room feel better on your skin, so you can turn the HVAC temperature up while retaining the same comfort level.

- **Always use efficient fans even when the air conditioner is on.** Moving air makes you feel better, so you can dial up the temperature in the room and get the same effect.

- **Keep fireplace dampers and flues tightly closed when not in use.** Not just seasonally, but all the time. If you're only using your fireplace on the weekends, close it up the rest of the time.

## Structural changes: Windows and roofs

You may benefit from investing in new windows, or a new roof. Windows have the potential to add a great deal of beauty to your home, so investing in them can be functional (if you buy ones that operate better than your old ones), energy efficient (many new materials and manufacturing techniques are available), and aesthetic (windows are generally the visual focus of a room, or at least they have that potential). Roofs periodically need to be changed, but you should consider installing an energy-efficient roof when your old roof poops out. A light colored roof reflects a lot of sunlight and therefore reduces the air-conditioning load. On the flip side, a dark, porous roof decreases your heating load. Which type is best for you depends on your climate.

## r-value

*r-value* is a complex technical term used in the industry to denote the level of insulative properties an insulation material offers. In general, the higher the r-value, the better the insulation. When you buy insulation, your dealer will be able to tell you what your local code requires for new homes in your area. You can decide whether you want to spend extra money for a higher r-value. You don't need to increase your r-value in the entire house to get some benefit of the extra investment. The family room is the most important room in the house because it's where your family spends the most time

### Using high-tech storm windows

Many old houses come complete with an entire set of storm windows, and putting them up and taking them down is a ritual of seasonal change. Basically, they accomplish the same thing as double-pane glass.

A relatively inexpensive thing you can do is have new, high-tech glass put into the storm window frames. You can get really good performance. However, keep in mind that old houses with storm windows usually don't have very good insulation in the walls. Simply improving the r-value (see sidebar) of the windows may be only a drop in the bucket.

### Getting new windows

Changing windows is expensive, but it can have a profound effect. You can get all kinds of different glass. Some is designed for cold climates — it lets in sunlight but insulates well for heat. Glass for hot climates blocks off sunlight, particularly in certain parts of the spectrum such as the UV (ultraviolet), which can cause damage and bleaching of fabrics inside the house.

Double-pane glass is the norm now. You can get triple pane, but it's expensive (and even more expensive when Johnny hits a baseball through it).

You don't need to change all the glass in your house. The family room or living room is the most important. Plus when you change a window, it's a major decorative feature. You can change the big window in your main living area, and if you change nothing else you've dramatically altered the character in that room. If you put in an energy-efficient window, with appropriate window coverings, you can also change the way the room feels. For example, direct sunlight in the summer may not be hot to a thermometer, but it feels hot on your skin, and that's what really matters.

Window contractors are in the phone book. Call one that deals with a number of different brands. A lot of them carry only one brand, and you're not likely to get a fair and balanced appraisal of all the options. Many window suppliers work with contractors a majority of the time. Sometimes these companies are not really enthusiastic about working with onesy-twosey private accounts. Look around for the right supplier because the prices vary quite a bit.

## Swimming pools and hot tubs

Swimming pools lose a lot of heat from the wind. If you simply install some low bushes on the windward side, your yard will be prettier, and you'll use less energy heating your pool. If you don't heat your pool, it'll stay warmer, and your season will be longer. You can even get drought-resistant bushes that you don't have to water, ever.

Most people with swimming pools run the filters for much longer than they need to. Two or three hours a day for a swimming pool usually works just fine. Try it and see whether it changes things appreciably. Pool pumps take a lot of power; the less you can use them, the better. If you're on a TOU (time of use) rate schedule, in which the price of power is higher during the afternoon hours than the morning and evening hours (head to Chapter 6 for a more detailed explanation), run your pool pumps at night.

Old hot tubs use so much energy it may be a sin. You can improve the situation with a good cover. Look in the phone book for suppliers; they're all over the place. Get a good cover; weather takes a toll, and the cheap ones crack and crumble after a few years, plus they're not as well insulated. Cheap covers are not inexpensive; they're just cheap.

If you have a redwood tub, get rid of it and buy a new energy-efficient model. They're very nice. The one I have came with preprogrammed filter cycles. I turn them down to half-usage, and the water is just as clean as before I made the change, plus I don't have to change filters nearly as often. If your filters are clogged, the pump is working too hard. Loading your pump (making it work harder than it needs to) costs a lot more than cleaning the filters periodically. Of course, I'll admit that watching TV is more fun than cleaning a filter.

If you have an old fiberglass style hot tub with meager insulation, you can make a nice do-it-yourself project by squirting expandable foam insulation around the outside of the tub's walls, beneath the outer cover. Have fun with this stuff. Make sure to wear your grungiest clothes, preferably ones you don't mind throwing away when you're done. Realistically, if you insulate an old tub you can expect the heating costs to halve.

# Window Covers: Blinds, Awnings, and Shades

You can achieve better insulative properties with your windows by simply covering them with appropriate materials (sun reflectors, insulation webs, and colored fabrics), and making it a habit to open and close blinds and curtains at certain times of the day, depending on the weather and time of year.

## Blinds

Blinds can be great insulators. The honeycomb variety work well in both hot and cold weather. Close them at night or when you're not home. They can also be excellent absorbers or reflectors of sunlight, depending on what type you get. In cold climates, you want blinds to absorb sunlight and warm up the room. You can do the same if you open the blinds, but then you won't be insulating. There are a wide range of blind styles, and it's impossible to get into details here. Big box hardware stores all carry a number of brands and have very good samples you can get your hands on. Prices vary from $20 per window to more than $300 per window.

On the other hand, if your main problem is heat in the summer, you want a blind that will reflect sunlight as much as possible. If you simply put a blind inside a window to keep sunlight out of the room, it'll absorb a lot of heat, but the heat will still be inside the room. You can put blinds over windows on the outside, which works much better in hot weather because the sunlight is stopped outside. Of course the blind gets really hot, but who cares? Go to your big-box hardware store to see the various options. The best bet, from my experience, is the solar screen variety because you can see through the material (sunlight is attenuated 80 percent or more, so the view is darker), but you can still see the view outside. (See Chapter 9 for more on blinds.)

If you have a big picture window that lets a lot of sunlight in on hot days, you can hang a cheap, roll-up blind on the outside. Get one that's solid so that it completely cuts off all light. If you can, leave an air gap of 6 inches or more between the window and the blind. Yes, the room will be dark and creepy, but if you're at work for the day, it won't matter, and the room will feel a lot better when you get home. Buy a cheap one because it probably won't last more than a season or two hanging outside in the direct sunlight. Take it down and store it when autumn arrives.

# *Awnings*

Awnings are great light shades as well. You can buy them with a variety of slat structures that accomplish a number of different functions (see Figure 3-2).

The Venetian awning allows sunlight in the winter, when the sun is low in the sky, and blocks sunlight in the summer, when the sun is high. Venetians also allow you to see out of the top of your window, and the effect is much more open than the hood awning.

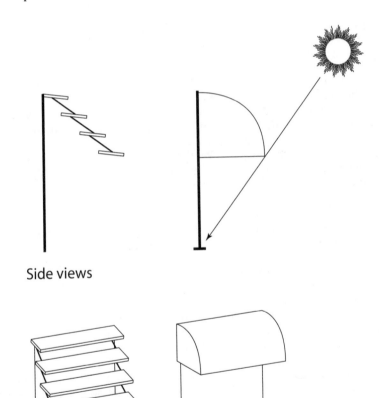

Side views

**Figure 3-2:**
Awning
styles.

Venetian awning          Hood awning

Hood awnings are common because they're more decorative. They can be made of nice fabrics that match the house. They work better in rainy climates because they keep cold water off your windows. You can lose a lot of heat by water-based convection from rain.

For the do-it-yourselfers, you can build very nice-looking wood slat awnings that can be attached to most surfaces. You can buy designs over the Internet, but even more fun is designing them yourself. Calculate the best angles by creating sun charts (see Chapter 5). You can paint them to match your house, and you can use materials that last a long time. You can even devise an adjustment scheme that allows you to raise or lower the pitch, depending on conditions.

## Shades and screens

Solar shade cloth works very nicely, and you can remove it in the summer. *Solar shade cloth* is similar to conventional screens, but it's thicker and has a specially designed mesh to inhibit infrared and ultraviolet light, neither of which do you any good in your house other than heating (since you can't see either wavelength). Solar shade cloth also has more integrity, and lasts longer, so the added expense isn't as bad as it may seem. Newer fabrics allow light to pass through without too much blur. On hot summer days, a lot less light is desirable — even if the room temperature is the same, the room feels cooler with less sunlight. Solar screens work better than sunlight-inhibiting blinds because they're on the outside, and the heat isn't dissipated on the inside as it is with blinds.

A good do-it-yourself project is to upgrade your existing screens from the cheap bug screens that are so common, to the solar shade cloth type materials. You don't need to change all the windows, just the ones where you want to inhibit sunlight. Big picture windows facing south are the best candidates. You can get screen at most hardware stores, but you can get the better performing stuff online or in specialized solar catalogs. It's a very good cost/benefit tradeoff, perhaps one of the best.

*Automatic retractable shades* are available from specialty suppliers. In my area, the big expensive houses generally come with electrically retractable shades mounted outside on roller tracks. Many of these houses have two floors of continuous windows facing west. On hot summer days, the sunlight can be murder, and without a shade the house would be uninhabitable. Plus in the winter, they can act as insulators.

Retractable shades are also commonly used overhead for patios and porches, where late afternoon and evening sunlight can prevent you from using the area. In this case, leave them up all day, or your floor will heat up.

Let the shade heat up instead, because any breeze will quickly cool it off — not so with concrete or tile floors, which hold heat for a long time. In the winter, don't use the shade because you want the floor to heat up and retain heat, which will dissipate back into the area when the sun goes down.

# Optimizing Operating Schedules

Most people set their thermostats to the desired temperature and then leave it be forevermore. But if you adjust your home's temperature at different times of the day, you can achieve cost savings without any decrease in standard of living. This advice is true in both summer and winter, and particularly when nobody is home during daylight hours.

## Controlling heating and cooling

Most people adjust their thermostats manually. You set the temperature, and your heating, ventilation, and air-conditioning system (HVAC) decides how to get the house there. Conservation calls for setting the temperature at 68°F in the winter and 82°F in the summer.

You can shut off your air conditioner at night, and your house will likely stay cool enough to be comfortable. If the temperature is hotter outside than in, don't open your windows. If vice versa, open up. You can also lower the thermostat temperature in the winter at night and sleep snug as a bug under a good comforter.

### Busting some popular myths

For the most part, homeowners try to minimize their use of the HVAC. Everybody understands it costs a lot. You can take further measures to reduce your costs, but first I need to dispel some common misconceptions:

- A lot of people think that if you turn your system off for a while, it works all that much harder when you turn it back on. This is not true. HVAC systems are either off or on. They don't work harder under any circumstances. They just work longer. So if you turn your system off, the energy you save during that time period will be equal to the energy you need to get your house back to normal temperature. If you own a heat pump, they work most efficiently when they're set to a temperature and then left on.

You don't save money by turning your thermostat off and on. You do save money by turning the temperature setting down, or in the summer, turning it up.

✔ A lot of people think that if you want your house to get to the set temperature faster, you turn the thermostat past the desired set point. For example, you come into a cold house in the winter and turn the HVAC on and then adjust the thermostat up to 90°F so things will heat up fast. The system doesn't work this way. The temperature will increase only so fast — and no faster.

You save money by turning your thermostat down as much as possible whenever you can (vice versa in the summer). If you can turn it down 15°F overnight, for eight hours, you can save 5 to 15 percent off your energy bill. Whenever you're gone for a few hours, turn your temperature down (in winter). Vice versa for summer air-conditioning season.

### *Programming your thermostat*

Programmable thermostats can save a lot of money. You can install them yourself. Ask at your local hardware store.

There is no universal standard for thermostat controls and if you wire up the wrong one, you may damage your HVAC. So make sure you get a thermostat that's compatible with your system.

Programmable thermostats accomplish the following:

✔ Make daily or weekly programs that turn your temperatures up and down without your attention

✔ Allow manual override so you can always set the temp where you want it, if only briefly

✔ Have an occupancy feature that can tell whether somebody's in the house; otherwise the temperature stays at standby

I highly recommend automatic controllers. Make sure to get one that has easy programming and good instructions written clearly on the unit itself; otherwise, you may be constantly searching for the manual. Also get one that has a clear, legible, and commonsensical display. It's amazing how counterintuitive some thermostats are. Who designs these things, anyway?

# Opening and closing windows and doors

In the same way that you can save by programming your HVAC system to operate at varying levels at different times of day or night, you can also achieve cost savings by strategically opening and closing windows and doors over the course of a day.

When you want breezes through your house, open windows and doors only on the windward and leeward sides of the house, and try to balance the openings. Don't just open every window you can. Experiment a little and you'll see the different results.

In addition, follow these tips:

✔ For best effect, use blinds and window coverings in conjunction with opening and closing windows. (See "Window Covers: Blinds, Awnings, and Shades," earlier in this chapter.)

✔ For cooling purposes, open windows that aren't in direct sunlight. Windows on the north are the best candidates.

✔ If you have multiple stories, opening windows on different floors can increase the flow of breezes. Experiment to see what works best.

✔ In the winter, open all blinds that are exposed to the sun. Let the sun shine into the room instead of striking the blind. Close the blinds when the outside is cold and lightless in order to insulate the window as much as possible.

✔ Use blinds and drapes wisely. Open them when you want sunlight and heat to come in. Close them when your air conditioner is running. And most people don't do this, but close them on cold nights to add an extra layer of insulation to the windows.

# Part II
# Understanding Solar — Just the Facts, Ma'am

"I don't know much about alternative energy sources, but I'll bet there's enough solar power being collected on those beach blankets to run my workshop for a month."

# In this part . . .

*B*efore you start getting your hands dirty with solar projects, you need to understand the nuts and bolts of the basic technologies behind solar equipment. You should know how best to mount solar collectors to get the most sunshine, and you need to be able to decide which solar investment will get you the best return on your hard-earned investment dollars. In this part, I show you how to analyze your house and decide where equipment should be mounted, and I provide some guidance on which direction you should go to find your best solar projects, given your funding and skill with tools.

# Chapter 4

# Working with Solar Basics

*W*hen you understand the basics of radiation, you're in a better position to decide on the optimum solar-energy system to install in your home. You don't have to rely on a salesperson's wild claims and potentially suffer an avoidable disappointment when things don't pan out as advertised. Knowledge is power, and *power* is what this book is all about. It can take many forms.

Photovoltaics (PV) have proliferated in the last decade due to many factors, the most influential of which is the marked increase in the government subsidies. In this chapter, I describe the main types of PV panels and their relative strengths and weaknesses. PV costs a lot of money, often as much as a new car, and it's important to understand the finer points of the options you'll face with when you choose to install a PV system. Of course, every PV contractor will claim its technology is superior, but you have to judge for yourself.

## Understanding the Nature of Light

The sun is a huge nuclear reactor that converts its own mass into light particles, or *photons*. Trillions of these photons fill Earth's atmosphere every second, and life on the planet relies on this energy.

Imagine snakes squirming frantically across a hot road. Their entire bodies wriggle in such a way that the motion pushes them forward. You can picture a photon precisely the same way, except the squirming, or *oscillating,* is much faster. And photons never rest; they're always moving at the speed of light. An important aspect of a photon is it's wavelength, which is just what it sounds like. In the case of the snake, the wavelength is the distance between two successive peaks in its wriggling motion. Larger snakes have longer wavelengths, and vice versa. Larger snakes also wriggle much slower than smaller snakes, and this is called the frequency. Wavelength and frequency are inversely related; you'll encounter both terms in the literature, but they basically just mean the same thing, except in mirror image.

Wavelength is important for solar systems because the physical components that make up a solar system respond differently to different wavelengths. For example, the sun puts out far more infra-red radiation than visible radiation, although we humans don't see any infra-red at all. In some solar applications, it's the infra-red that is the most important; in other applications we may only be interested in visible light.

When you focus your eyesight on an object, you see the light that has reflected off that object. So when you see a green plant, the plant has absorbed all the wavelengths besides green. When you see white, you're seeing all the colors at the same time. A white object generally doesn't absorb any colors of light at all — it reflects them. On the other hand, a black object absorbs everything. This property is very useful for solar projects because you're interested in creating equipment that absorbs as much sunlight as possible.

The peak of the spectrum occurs in the visible region. Below visible (on the wavelength scale) is *ultraviolet* light, and above visible is *infrared*. Each photon has energy, and you see different colors because your eyes react to differences in energy. Red photons have less energy than blue, and infrared photons have much less energy than ultraviolet. From a physics standpoint, there's no difference between photons that are visible or invisible, aside from different energies, but your eyes discriminate and draw conclusions. Figure 4-1 shows the complete spectrum of sunlight.

# Getting off light: The sun's good behavior

Here are some facts about the sun:

- The sun's nuclear reaction consumes 4.2 million tons of mass per second. Not to worry, though. At this rate, the sun will last another 6 billion years, which is probably a lot longer than humankind will last, especially if we keep consuming energy the way we have been.

- The Earth intercepts only 2 billionths of the sun's total energy output. Humans are

but insignificant pipsqueaks in the grand scheme of things.

- The amount of sunlight that falls on the Earth's surface is 35,000 times the amount of total energy used by humans. That's a lot of sunlight to spare.

- At sea level on a clear day, 1 kWh of sunlight falls on a 1-square-meter area. That's enough power to run most appliances in your house.

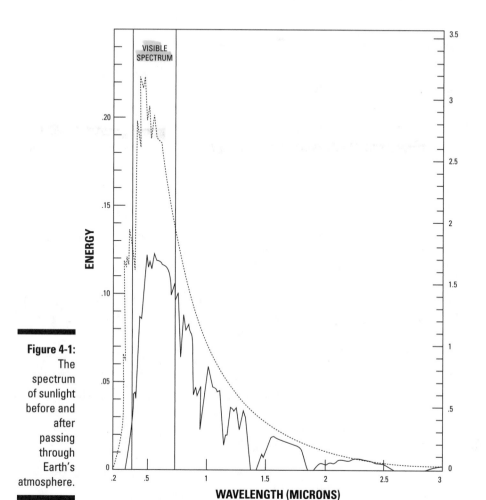

**Figure 4-1:**
The
spectrum
of sunlight
before and
after
passing
through
Earth's
atmosphere.

# Here Comes the Sun: Radiation Reacting with Matter

In the context of solar power, *radiation* simply means sunlight, the entire spectrum. When engineering a solar system, the interaction between sunlight and the various materials that comprise the system is of utmost concern. Without the proper interaction, the system will be inefficient and may not work the way it's supposed to.

# A bit of light interaction

Radiation reacts with matter in a number of ways, all of which can play a role in your solar-power system:

- ✔ **Absorption:** When a photon is absorbed, it's energy is changed into a different form, either heat or electrical energy. Every solar system is primarily concerned with absorption, for until that happens there is no capture of energy.

- ✔ **Transmission:** When light simply goes right through a medium, it's called transmission. Windows are good transmitters, and the atmosphere is a rather good transmitter when the air's clear and smog free.

- ✔ **Scattering:** Scattering occurs when photons interact with molecules, the result being that a "secondary" form of radiation emits in many directions at once. Scattering is responsible for such diverse phenomenon as rainbows and the dusk/dawn reddening of the sky.

  In the atmosphere, light scatters off individual molecules, and scattering occurs much more readily with blue light than red. Hence, when you look up into the sky, you see blue because blue light is bouncing around all over the place. On the other hand, when you see light that has traveled through a lot of atmosphere (in the evenings and mornings) it often appears red because that wavelength passes through the atmosphere the most easily.

- ✔ **Reflection:** Mirrors are reflectors. In general, shiny, hard surfaces are much better reflectors than porous, dark surfaces. Reflection differs from scattering in that the direction of a reflection is a mirror image of the incident (the original) photon.

Some mediums pass radiation, but also insulate heat very well. This idea is useful with solar collectors because you want sunlight to enter your system, but after it's converted into heat, you don't want that to get back out. Glass impregnated with iron silicon transmits wide spectrums of light and also insulates heat very well.

When a medium doesn't transmit or reflect all wavelengths equally, it's called a *filter*. People can make windows that pass visible light very well but reflect ultraviolet. Sunglasses are a perfect example of filters. By selectively filtering certain wavelengths, for example, ultra-violet, visibility can be enhanced while cutting out heat. Filter performance also comes into play when designing glazing for solar water heating collectors (glazing is the material that covers the collector and allows sunlight to enter the system; see Chapter 12). In this case, it is desirable to allow the maximum amount of sunlight to enter the system, and so the filter properties are centered where the most sunlight is, namely the visible and near infra-red regions of the spectrum.

All materials are a combination of reflectors, transmitters, and absorbers. For example, windows transmit only around 85 percent of the light that strikes them; they reflect the rest. Likewise, Earth's atmosphere reflects, absorbs, and transmits, and each acts as a filter because each effect depends on the wavelength of the incident light (*incident* just means the light that's coming into a surface). The windows in your home, while they may not seem like it, filter out a lot of sunlight. When people mess around with the composition of the atmosphere, for example by loading it with carbon dioxide or other pollutant molecules, they end up inadvertently changing weather patterns. That's the gist of global warming, which I touch on in the next section.

## *Look! Up in the sky! Light through the atmosphere*

Figure 4-1 (earlier in this chapter) shows the difference between the spectrum of sunlight that hits Earth's upper atmosphere and the spectrum that reaches ground level. Under normal conditions, as much as 35 percent of solar radiation is reflected back into space or absorbed before it hits the surface of the Earth. In the presence of clouds or pollution, the reflection and absorption can be much greater.

Water vapor, carbon dioxide, and ozone absorb 10 to 15 percent of sunlight. In the upper atmosphere, *ozone* (three oxygen atoms bonded together, or $O_3$) filters out virtually all the ultraviolet radiation (short wavelength radiation — very energetic. Note in Figure 4-1 that plenty of ultraviolet light comes from the sun, but almost none of it actually reaches the Earth's surface. If it were to reach the planet's surface, life would drastically change.

A number of years ago, people were greatly concerned about the hole in the ozone layer. But this concern was partially allayed with the banning of certain aerosols and refrigerants that contained ozone-depleting compounds called chlorofluorocarbons (CFCs), commonly known as Freon. Here's a good example of how environmental policy actually works!

## The greenhouse effect

Some materials are very good at transmitting radiation while at the same time insulating for heat. For example, as Earth's atmosphere fills up with carbon dioxide, the air still passes along the same amount of radiation, but the atmosphere becomes better at trapping heat. The net effect is that the Earth is heating up more than it would under completely natural circumstances.

Glass (which you find in actual greenhouses) likewise transmits radiation and insulates at the same time. To give you an idea of how profound the effect is, consider that a car sitting in the sun heats up because of the same greenhouse effect. And everyone knows just how miserably hot a car can get.

In addition, atmospheric depth makes all the difference. At noon, when the sun is directly overhead, sunlight passes through much less atmosphere than at dawn or dusk (see Figure 4-2).

You can see why the middle of the day is hottest. It's not because, as some people believe, the sun is closer. Rather, the sunlight passes through much less atmosphere. The Earth is tilted on its axis, –23° to be precise. Figure 4-3 shows what happens as Earth rotates.

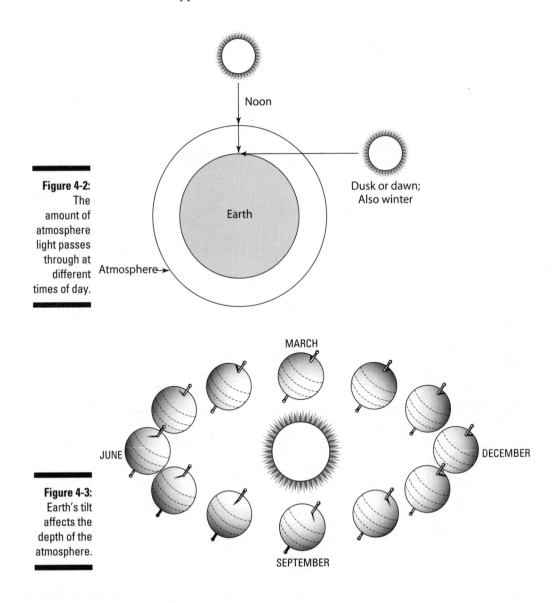

**Figure 4-2:** The amount of atmosphere light passes through at different times of day.

**Figure 4-3:** Earth's tilt affects the depth of the atmosphere.

The sun seems hotter in the summer than in the winter, but that's not really true. The sun doesn't change; it's the air and the position of the Earth that changes. Less sunlight energy reaches the Earth's surface in the winter for the same reason that you get less solar energy at dawn and dusk.

# Using Sunlight Converted into Heat

When a photon (light particle) strikes a surface that absorbs light, the photon's energy is transformed into heat. *Heat,* or *thermal energy,* is simply motion. A hot surface has a lot of molecular motion, and hot objects can burn you because they pass their excessive energy right into your skin.

The following sections explain the basics of heat transfer and how that may influence what goes into your solar collector.

## On the move: Modes of heat transfer

Many solar applications require sunlight to be converted into heat. Then you want to transfer that heat and change it into a usable form, usually in a location removed from where the heat is captured. Heat can move from one spot to another in several ways:

✔ **Conduction:** In conduction, energy is transferred between molecules within a substance. Only heat moves, not the molecules. (Well, molecules are always moving, but the molecules tend to vibrate in place rather than switch places.) Heat always moves from a hotter surface to a cooler surface (technically speaking, heat moves in both directions, but statistically the most movement occurs from hotter to cooler). Copper is an excellent conductor of heat. In fact, most good conductors of electricity are also good heat conductors. Glass is a poor conductor of heat and electricity and is therefore considered an insulator.

✔ **Convection:** Convection is the transfer of heat between a fixed, rigid surface and a moving fluid in contact with that surface, or the transfer of heat as molecules move from one point to another within the fluid. A *fluid* is any substance that can flow, so air is also considered a fluid. As opposed to conduction, the hot molecules themselves move and switch places.

Water flowing over a hot surface becomes hot. In a tank of water, the hotter molecules expand and weigh less and move upward while gravity draws the colder, heavier molecules down. This heat exchange is called *thermosiphon,* and it's a type of convection.

✔ **Radiation:** A hot object emits infrared radiation. The sun is a perfect radiator. Fires radiate heat, a lot of which you can see. A really hot fire glows white; all wavelengths in the spectrum are present. As the fire cools down, the color changes to orange, then to red, and finally, when you can't see the embers anymore but can still feel the heat, the radiation is entirely infrared. The colors reflect the energy content of the fire.

In Chapter 10, I get into detail on how solar heating systems make use of these properties. In fact, every engineered system must take into account heating and cooling concerns because there is always a transfer of energy from one part of a system to another. If there weren't a transfer of energy, nothing would be happening, and it would be a political system instead of an engineered system.

## Keeping the heat where you want it

Solar heat collectors are designed to collect as much solar radiation as possible. Not only do you want to maximize surface area, but you also want to orient the collector in a way that maximizes the amount of solar radiation gathered over the course of a day. Solar heat collectors also ought to do the following, all of which affect what kind of materials go where in your solar collector:

✔ **Convert that radiation into heat as efficiently as possible.** This conversion almost always entails the use of a black surface, which absorbs radiation most efficiently.

✔ **Transfer the heat into a usable medium.** In most cases, you want to heat water. If you collect the heat in a black tube, you can simply run water through that tube, and the heat naturally transfers into the water. Other fluids, such as antifreeze fluid (which doesn't freeze up in winter) may also be used.

To move heat effectively, a material needs to be a good conductor (or as physicists like to say, it needs to have *high thermal conductivity*).

✔ **Insulate to prevent heat loss.** For insulators, you want to use materials that make for poor conductors. Converting radiation into heat in the wintertime is easy, but then the cold weather and wind simply steal the heat right back. In order to prevent this loss, people insulate the collector from the outside world with materials, such as fiberglass insulation, or double-pane glass windows.

✔ **Store a sufficient amount of heat.** *Heat capacity* is a measure of how much heat energy (in British thermal units, or Btus) is required to raise the temperature of a given volume of a material by one degree Fahrenheit. To be effective at storing heat, a material needs a high heat capacity. Table 4-1 shows some materials of interest.

| Table 4-1 | Heat Capacity of Common Materials |
|-----------|-----------------------------------|
| *Material* | *Heat Capacity (Btus per Cubic Foot-°F)* |
| Air | 0.018 |
| Plastic | 0.57 |
| Wool, fabric | 2.2 |
| Concrete | 22 |
| Brick | 25 |
| Wood, oak | 26.8 |
| Steel | 59 |
| Water | 62.4 |
| Copper | 78 |

So the ideal material for storing and moving heat needs to have a high heat capacity and also has to be a good conductor. Although wood and concrete store a lot of heat, you can't move it into or out of these materials very easily. Steel is very good, but it rusts. Of course, water is the cheapest, so it's used to store and transfer heat, but the problem is that it doesn't absorb radiation because it's clear. Copper painted black absorbs, stores, and conducts heat extremely well, so copper is used extensively in solar heat collectors (the price is also relatively low, in comparison to other metals that could be used).

# Converting Sunlight into Electricity: Photovoltaic Cells

Most people are familiar with photovoltaic cells (PV cells). You see them on calculators, the kind that don't require batteries. You see PV panels on people's roofs and on businesses and schools and government buildings. Remote telephones along the interstates have PV panels poised overhead.

A standard PV cell, shown in Figure 4-4, is a thin semiconductor sandwich, with two layers of highly purified silicon. A *semiconductor* is a crystal that is constructed in such a way as to yield certain specific properties that can be used in electrical circuits. Transistors are made of semiconductors. Microprocessors are as well. In fact, the entire field of modern electronics uses various forms of semiconductors, so the implications are worldly. If you connect an electrical circuit to your semiconductor sandwich when sunlight ? Photovoltaic arrays are nothing more than huge matrices of interconnected semiconductor sandwiches. The color appears shiny blue. Usable PV systems are comprised of all sorts of equipment that protects the user from electrical shock, stores the electricity in battery banks, and converts the direct current (DC) into alternating current (AC), which is what people use in their houses. But at the heart of each system is a simple conversion process like that shown in Figure 4-4.

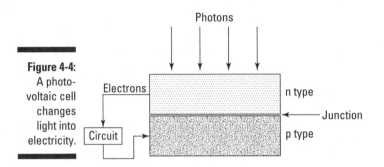

**Figure 4-4:** A photo-voltaic cell changes light into electricity.

# Understanding Photovoltaics in More Detail

For the most part, if you're considering installing a solar PV system on your home, you don't need to know the nitty-grittys of how the PV cells work. Your solar contractor knows the details (they should, at any rate), and they know which types of panels to use in a given application.

But PV systems cost a lot of money, and customers are generally interested in knowing as much as possible about the details. The more you understand, the better your own decision-making process will be. You'll also be in a better position to judge whether your PV contractor passes muster. In this section, I describe some of the more important aspects of PV cells, with an eye toward giving you the information you need the most. I try to leave out the gory stuff as much as possible, and in this subject there's a lot of the gory stuff.

## The basics of every PV cell

A basic cell is around 1/100th of an inch thick, with a wide range of surface areas. A typical life span is over 25 years, and there are cells in place that have produced electricity reliably for over 40 years. The more difficult technical problem is encapsulating the cells into a weather-resistant housing. It's the coatings and framing structures that influence the ultimate lifetime of a PV cell. Very few technologies are called upon to withstand the harsh conditions an average PV system must withstand.

A module is an assembly of individual cells, connected in series and parallel arrangements designed to yield optimum performance. Figure 4-5 shows the difference between a series and parallel connection:

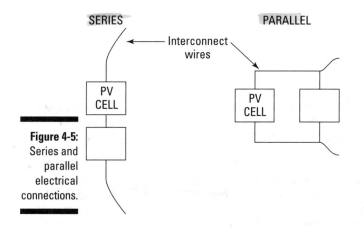

**Figure 4-5:**
Series and
parallel
electrical
connections.

In a series connection, voltage is additive, while in a parallel connection, current is additive.

A typical PV cell produces around half a volt of electrical output. When 36 PV cells are connected in series, the result is an 18 volt module. The term *module* is used interchangeably with the term *panel* — a module or panel is a number of individual cells interconnected and housed into a finished product. A typical PV module in a residential application measures around 2.5 feet by 5 feet, with a color either bluish or black. Frames are either aluminum colored or black, with the latter being the overwhelming choice of most homeowners these days (they simply look better).

It's possible to achieve a wide range of voltage and current outputs, depending on how the individual cells are connected together. The amount of power a module can produce is a function of the total surface area, as well as the amount of sunlight that strikes the module.

Typical modules are rectangular in shape, and available in a variety of sizes and configurations. Small modules output less than a single watt of power (the kind of modules used in hand-held calculators), while a typical residential module produces around 200 watts of power, more or less.

## *Looking at different module types*

Individual silicon wafers used to manufacture PV cells are embedded with metallic contacts (wires). The cells are coated with an anti-reflection material so that the maximum amount of sunlight is absorbed into each cell.

Modules are characterized by:

    ✔ Cell material, or the type of silicon process that is used

    ✔ Glazing material

    ✔ Frame and electrical connections

The most important feature of a cell is the composition of the silicon structure. Single crystalline cells may be cast into an ingot of multiple crystals (referred to as *polycrystalline*). Or the crystalline materials may also be deposited as a thin-film, which is referred to as *amorphous silicon*.

Single crystalline cells are more efficient than polycrystalline due to the fact that in polycrystalline cells, inter-grain boundaries introduce resistance to current flow (which consumes energy). Amorphous silicon is much less expensive to manufacture, but it's only around half as efficient in converting sunlight into useable electrical energy. In practical terms, this means that an amorphous system requires twice the surface area to output the same amount of power. Depending on how much surface area is available, this may or may not be a problem. In most residential applications, suitable roof space is limited, so efficiency is an important factor.

Here's an approximation of the efficiencies and cost of the different PV technologies:

    ✔ Monocrystalline can achieve 15 to 20 percent efficiency, with a cost of around $3.48 per watt.

    ✔ Multicrystalline achieves between 12 and 15 percent efficiency at a cost of around $3.29 per watt.

    ✔ Thin-film technology can only achieve 4 to 14 percent efficiency, but it costs less than $2.50 per watt, and this price is coming down by leaps and bounds.

There are over 50 PV module manufacturers at this writing, and new plants are being built around the world at a high rate. Many new materials and processes are being developed, and it's impossible to summarize these here.

## Analyzing module performance

There are four predominant factors that determine a PV module's performance: I-V curve, amount of incident sunlight, cell temperature, and shading, all explained in the following sections.

### I-V curve

Modules are characterized by what's referred to as an *I-V curve,* or a current versus voltage curve. The technical details are beyond the scope of this book, but suffice to say that a module needs to be connected to a load (which is where the electrical power is going) that matches the module's characteristics as closely as possible. Because sunlight intensity varies (meaning the optimum load also varies), it's impossible to operate a PV module optimally under all conditions. A solar PV contractor uses sophisticated algorithms for determining the best load, averaged over all the expected conditions.

### Incident sunlight

The amount of sunlight incident on a module directly determines how much power output the module will be capable of (go to Chapter 5 to get into more details about how to best locate a module).

### Cell temperature

Cell temperature affects the power output of a PV module more than you might expect. In general, heat causes inefficiencies in electronics. The output of a solar PV module goes down with increasing temperature, which is kind of ironic if you think about it since the goal is to maximize the amount of sunlight on a panel. But with more sunlight comes more heat, and as a module heats, it becomes less efficient. In practical terms, you can get more output from your solar PV system on a partly sunny, cold day than you may get on a very sunny, very hot day. A good breeze also helps to cool a module, so the best conditions are cool, breezy, sunny days.

The fact that solar production decreases with temperature causes many new solar customers concern. They intuitively expect their outputs to track the intensity of the sun, but because on really hot, sunny days they end up seeing their system outputting less, so they call their contractor and claim something's wrong. A good contractor will explain this effect to you up-front, so you're not surprised.

### Shading

Shading is the bane of all solar PV systems, some more than others. No surprise there, but the magnitude of the effect is more than you'd probably expect: A panel with only 5 percent shading may experience 50 percent reduction in power output. Therefore, positioning panels so that the shade is reduced as much as possible is extremely important. But in a typical application, some shading is inevitable. That's why the goal is to optimize module placement by taking into consideration all the conditions that may be expected over the course of a day and a year. Given a choice, shade in the morning is better than shade in the afternoon.

## Shining a light on PV arrays

An array consists of two or more panels connected together to achieve a particular voltage and current, given the ambient conditions. In a typical residential application, anywhere between 14 and 40 modules are used to comprise a complete system. It's best to use the same type of modules for the entire array.

Problems arise — it's very difficult to achieve optimum performance — when dissimilar modules are connected together.

Arrays may be wired in a wide range of series/parallel combinations, and each different option results in a different system performance. Good contractors use sophisticated algorithms to determine the most optimum wiring configurations. If you do it yourself, you need to understand how the various array configurations interact with the inverter (more on this in Chapter 16). The best advice is to let the pros design your system, even if you decide you want to install it yourself.

# Looking into the Future

The solar PV business is constantly evolving, and a wide range of new technologies are being developed. Higher efficiencies are desirable, but because cost is usually the driver, lower cost technologies are proliferating more quickly than high-efficiency technologies.

Large scale solar farms, often located out in the middle of nowhere, use low-cost thin-film technologies because they have no need for high-efficiency. In this situation, cost is the ultimate driver, not minimization of surface area. When utilities purchase a high-output solar farm, residential customers inevitably benefit because of economies of scale. The more solar PV built in the world, the lower the cost.

The development curve that PV manufacturing is following is similar to that of the semiconductor industry back in the 1960s. Prices plummeted while quality increased dramatically. A lot of the manufacturing equipment used to make PV cells is similar to that used to make microprocessors, except on a much larger scale.

The PV industry grew over 100 percent from 2007 to 2008, and projections are for this growth rate to continue. Ultimately, economies of scale and increased competition will force module prices down. But because many factors affect the price of a solar PV system, the price of solar PV residential systems may not fall in the same way.

Solar PV systems ultimately compete with the existing power grid. When solar power is cheaper than grid power, customers install solar systems. As grid power prices continue to rise, expect the price of solar power to also rise because the solar power industry can charge more for its equipment and still be competitive with grid power.

# Chapter 5

# Evaluating Your Solar Potential

*D*epending on where you live geographically — and the orientation and exposure of your particular house or business — you get more or less usable sunshine. Even within small, localized regions, weather patterns vary due to topography and landscape details like trees and ponds. So two identical solar systems separated by a few miles, but otherwise built and operated identically, may yield different energy outputs averaged over a period of time.

In this chapter, I explain the factors that allow you to optimize the orientation and location of solar systems at your home. This information can give you a head start in making important decisions, such as which type of system will work best for you, how big a system you will need, and how much potential, in relative terms, you can expect to achieve. Above all, this chapter gives you an idea how to evaluate the relative economics of a solar investment, the details of which I get into in Chapter 6, on payback analysis.

# Mother Nature in Your Neck of the Woods: Climate

*Weather* is today's phenomenon, but *climate* is a description of the general weather patterns over a long period of time. It may be cold and rainy in Los Angeles today, and that's the weather, but the climate is warm and temperate. Good solar designers assess climate particulars to enhance system performance. Climate includes elements such as temperature, precipitation, and wind speed, among other things. Here's a look at how climate can affect your solar system:

   ✔ **Sunlight:** Climate dictates how much sunlight you can expect annually. The map in Figure 5-1 shows the average number of hours per day of sunshine in the United States and Canada. That the Southwest gets the most sunshine per day — and that Canada and the northern states get the least — should come as no surprise. The sun is higher in the sky in the southern states, so the days are longer.

✔ **Snowfall:** You want to locate your panels so they avoid being inundated with heavy layers of snow. For example, some locations on your roof will experience very shallow snow buildups as compared to other parts of your roof. Also, some parts of your roof may be warmer than others due to proximity to heaters, exhausts, chimneys, and so on.

✔ **Cloud cover:** If you're living in a cloudy region, you still have solar energy, and it's generally diffused (spread out). As a result, collector panel orientation isn't so critical because light will be coming in at many different angles rather than just directly overhead from the sun. Ultimately, cloudy regions provide less sunshine, and solar systems are harder to justify. But that doesn't necessarily mean that solar is uneconomical, so if you live with a lot of clouds, don't despair.

✔ **Smog:** Air pollution and smog affect the amount of sunlight you can expect to receive. If you do live in an area with heavy air pollution, expect less system output over an extended period of time. Smog and air pollution also affect the spectrum of solar radiation that you will receive (In Chapter 4, I describe filtering affects). This in turn may affect the type of solar system that you end up choosing. Also be aware that smog and acid rain can be very corrosive, and some materials have a tough time over the long haul, plastics, in particular.

✔ **Air density:** You get better solar exposure in the mountains than near sea level simply because the air is thinner and scatters less sunlight. You can make an approximate estimate of how clear your air is by simply observing how blue the sky is on a clear day. Thick air scatters more red light, and so the appearance of the sky is less blue and more white.

✔ **Temperature:** With PV systems (not solar water heaters), the lower the temperature, the happier the semiconductors, and the greater the output. You can get more system output on a cold, clear day than a sunny day. Heat also tends to warp metallic structures as they expand and contract from changes in temperature. Regions where it gets very hot during the day and very cool at night can be very tough on certain types of equipment, in particular those relying on a lot of connectors and frame structures.

✔ **Rainfall:** Wet, humid environments tend to cause corrosion in metals. Electrical connections are particularly susceptible, and they either fail entirely or their integrity is compromised, resulting in poor system performance. It's very important to seal equipment junctions properly (a task a good contractor can perform).

✔ **Frequent fog:** If you're living in an area that's foggy and misty in the morning (in the San Francisco Bay Area, for example) but the mist burns off into a clear sky by noon, you want to orient your solar panels more westward to optimize the amount of sunlight you can achieve over the course of a day. Fog also causes a lot of moisture related problems, such as corrosion.

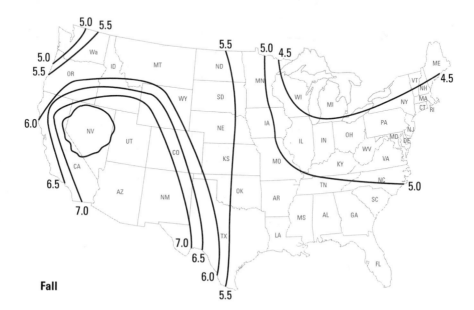

Fall

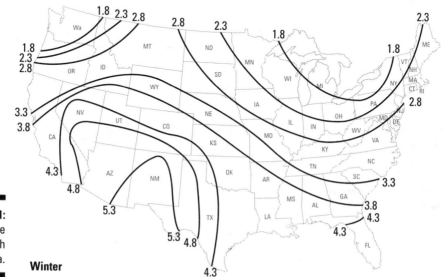

**Figure 5-1:**
Sunshine
in North
America.

Winter

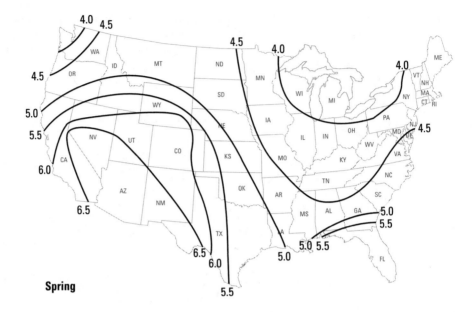

Spring

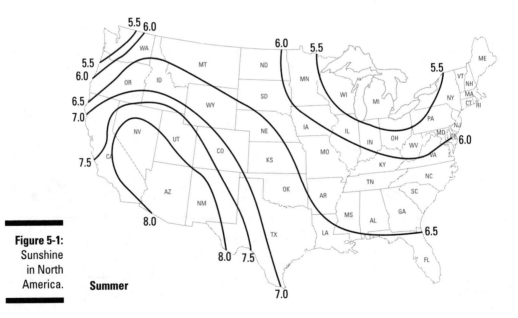

**Figure 5-1:**
Sunshine
in North
America.

Summer

✔ **Wind:** If you have a lot of wind, you need to consider where you mount your solar equipment for a couple of reasons:

- Wind can tear equipment off of its mounting hardware and result in expensive repairs, not to mention dangerous conditions. If you're in a windy climate, you need to make sure that you specify heavy-duty mounting equipment. Mounting schemes all have wind speed specifications; pay close attention because mishaps are expensive.

- Wind cools surfaces very efficiently (through convection — see Chapter 4 for details on heat transfer). If you have a swimming pool, one of your best bets is to install landscaping that breaks the rush of wind that will cool the pool water much more than your intuition would indicate. Solar water heating panels may heat the water very effectively, but it doesn't make much sense to install expensive solar panels without addressing wind cooling first, because adding a few bushes and trees will be much cheaper than adding extra collector capacity. Plus it looks nicer and is better for the environment.

You can obtain generalized information and maps about the nation's solar resources from the following:

**The National Renewable Energy Laboratory:** Located in Golden, Colorado, the NREL (www.nrel.gov) has a solar resources section on its Web site; follow the prompts to obtain a data log for your city, or a nearby large city. You can get an estimate of the BTU's/square meter/day you can expect. This data is valuable because it averages out all the factors that I explain in the previous bulleted list.

**National Climatic Data Center:** The NCDC (www.ncdc.noaa.gov), in Asheville, North Carolina, claims to have the world's largest archive of climate data.

## Considering climate and energy use

Your climate dictates how much energy you need and when you need it. In temperate climates, the requirement for heating and cooling is generally lower. In northern climates, you don't need to worry about cooling your house at all. Your problem is how to heat your house in the winter.

Part of assessing climate is what you want your solar system to do for you. If you have a cabin in northern Minnesota, you probably won't be there much in the wintertime. And then you'll heat it using renewable wood. In the summer, you don't need to cool, and all you want to do is obtain some nighttime lighting and run a small, efficient refrigerator. In this case, a modest, off-grid photovoltaic system with a battery backup can do the job. Even if you get only a few hours of sun each day, and it's low in the sky, you can still install a system that works economically for you.

# Plotting Your Sun Charts

*Sun charts* plot how much direct sunlight you can expect over the course of a day. They're an easy and intuitive way to visually display the movement of the sun across the sky. The following sections explain how to create your own sun charts and how to use them to evaluate the amount of sunshine you can expect to receive at your home.

## Charting out the basic path of the sun

The position of the sun may be plotted with two angles (*azimuth,* which is the angle from true south, and *elevation,* which is the angle from level, or in most cases the horizon), as shown in Figure 5-2.

Figure 5-3 shows you how to create a graph of the sun's passage over the course of a day. Imagine a sheet of graph paper wrapped around your house. As the day progresses, you make dots where the suns shines on the graph paper.

Figure 5-4 shows what your sun chart looks like if you plot the movement of the sun. The arc in the middle represents either spring or fall. All other paths lie somewhere between the two extremes, represented by summer and winter solstice, which are the longest and shortest days of the year (June 21 and December 21, respectively).

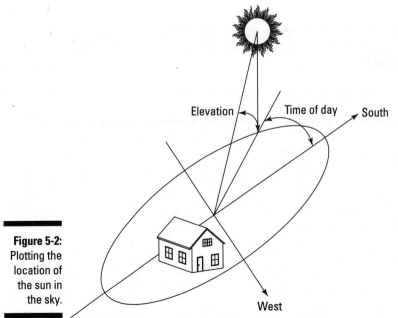

**Figure 5-2:**
Plotting the
location of
the sun in
the sky.

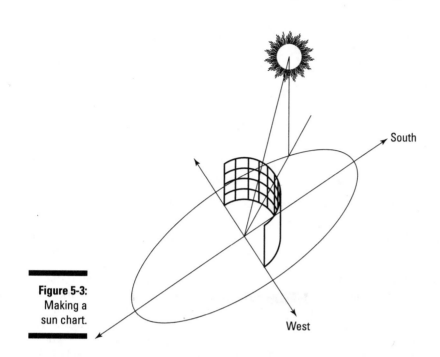

**Figure 5-3:**
Making a
sun chart.

South

West

 It's not necessarily true that the sun is highest in the sky at noon. In fact, it's rare. And it's also rare when dawn and dusk are symmetric around noon. But putting the sun at high noon and making dawn and dusk an equal distance from the noon hour lets me make my point more simply, so don't get nit picky. Nobody wants to date a techno-nerd, except maybe another techno-nerd.

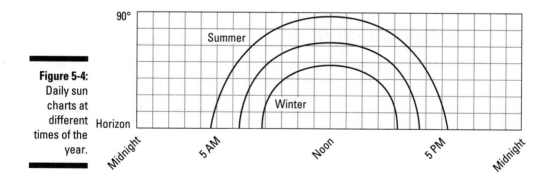

**Figure 5-4:**
Daily sun
charts at
different
times of the
year.

90°

Summer

Winter

Horizon

Midnight    5 AM    Noon    5 PM    Midnight

# Adding skyline effects

You can easily add a *skyline* (any impediments to direct sunlight, which include horizon, buildings, trees, towers, and so on) to your sun charts.

Trees, neighboring roofs, tall buildings, mountains — you can easily include each of these (see Figure 5-5).

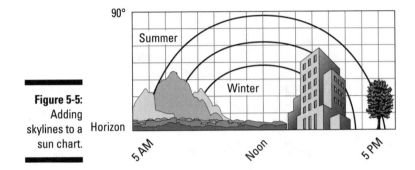

**Figure 5-5:**
Adding skylines to a sun chart.

You can either buy (very expensive, and you only need it once) or rent ($25 per week) a Solar Pathfinder (www.solarpathfinder.com), which works as follows: You stand at the site you want to measure, aim the device south (it has a compass), level it (it has a bubble level), and then read the shade reflections on a domed indicator. A chart under the dome shows hours and months when any bit of shade will be problematic. These devices are rather complex, but they're interesting because you can see how they incorporate all the ideas in this chapter. These devices also come in electronic versions, which can be plugged into a PC. With the provided software, you can explore all kinds of interesting solar panel siting scenarios. If you have a solar contractor come to your home to do a bid for a job, he may bring one of these devices and be happy to share the techno-nerd results with you.

## Noting sunlight intensity

In addition to how much direct sunlight you get, sunlight intensity is important. When the sun is lower in the sky, solar radiation must pass through more atmosphere, and it's therefore reduced by scattering and absorption. Figure 5-6, which is very similar in nature to a sun chart, shows how it works.

**Figure 5-6:**
Plotting sunlight intensity over the course of a day and season.

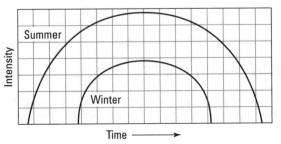

The sun is the most intense when it's directly overhead. And summer sunlight is much stronger than winter. You already know that — all you do now is make a plot to show what your intuition already understands.

Sunlight likewise changes along with the weather. Figure 5-7 shows what happens when the weather's cloudy or foggy. If your climate is often foggy or hazy in the morning, the charts show a very shallow curve on the left-hand side, and then when the fog burns off, the chart goes back up to normal.

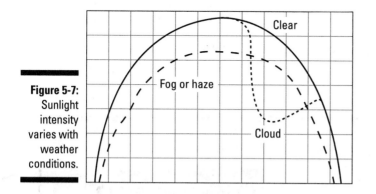

**Figure 5-7:**
Sunlight
intensity
varies with
weather
conditions.

# Collector Cross Sections

The final piece of the puzzle is *collector cross section.* Imagine a sheet of paper that you look at from different perspectives. If you set it on a table and look down, directly at it, it will look like a sheet of paper, nice and rectangular; you see the entire extent of the paper. If you look at it from the side, you will hardly see anything at all except a thin line. As you move your perspective around, the sheet grows bigger and smaller, and the rectangle becomes a parallelogram with odd angles. A fixed solar collector works on the same principle. The more surface area that is exposed to direct sunlight, the more output the collector is capable of.

How you orient your collectors with respect to the sun-chart angles is of critical importance. For example, if you mount your collectors facing north, you will obviously get less sunlight than if your collectors are facing south, simply due to cross-section issues. If you mount your collector facing toward the west, you will get afternoon and evening system output. In this section, I explore some results of cross-section engineering that are of practical use.

Pretend you're the sun, looking down on your own house. Figure 5-8 shows what a solar panel would look like from different vantages.

To get a better understanding of the relationship of the angles, set a business card in a reclined, fixed position on a table top and then move around from side to side, and up and down (mimicking the way the sun moves over the sky during the course of a day and over the seasons), noting how much area of the card you can see from different vantage points.

The difference in the sky between equinoxes (middle of winter and middle of summer) is 46°, which is considerable. Try it for yourself with the business card setup. If the angle is off by 25°, you still pick up 90 percent of available radiation (which means you see 90 percent of the full surface area). Off by 45°, and it goes down to 71 percent. Even at 80°, you still get 17 percent exposure.

Plus, there's a lot of scattering in the air. Even after the sun has set, some sunlight is still available. The spectrum changes quite a bit, but let's not get too techno-nerdy here.

# Earth: Going full tilt

The Earth is tilted on its axis 23.5°, and the following figure shows what happens to the position of the sun in the sky over the course of a year.

Regardless of where you live, the difference between the sun's peak angles in the sky from December to June is 46°. Regions closest to the poles experience seasons when the sun never shines, and six months later, seasons when the sun never goes below the horizon (known as *white night*).

The optimum elevation angle for your solar system depends on your latitude. In general, the optimum tilt angle is equal to your latitude, for this will ensure the maximum amount of sunlight exposure over the course of a year. Even better would be to manually change the elevation of your solar collectors over the course of a year to "follow" the sun's elevation in the sky.

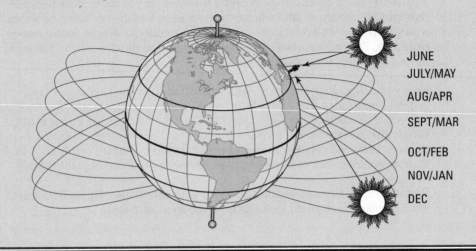

JUNE

JULY/MAY

AUG/APR

SEPT/MAR

OCT/FEB

NOV/JAN

DEC

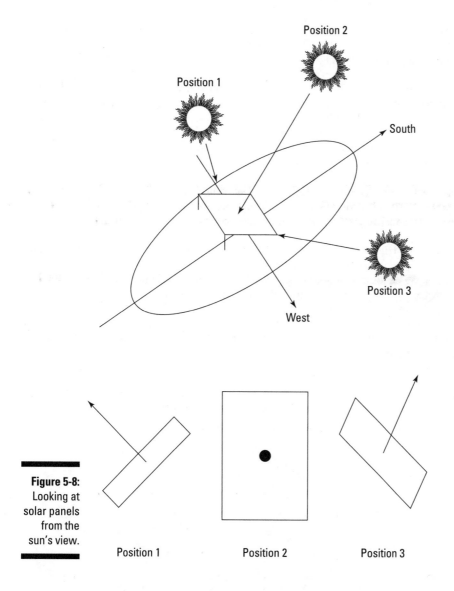

Position 2

Position 1

South

West

Position 3

**Figure 5-8:**
Looking at
solar panels
from the
sun's view.

Position 1          Position 2          Position 3

# Mounting Your Collectors Optimally

Figure 5-9 shows the usual options for mounting collectors around your house. You always have a number of choices for mounting, and the best choice depends not only on maximizing exposure over the course of a year, but also on cost and practicality. County codes require panels to withstand very high wind speeds, and conforming to this requirement can sometimes mean thousands of dollars in engineering and equipment. When all is said and done, simplicity reigns.

In most cases, you may not have much choice. If your roof faces southwest and its pitch is 45 °, that's how you'll end up mounting, unless you want to get into some really odd-looking and expensive mounting racks. Appearance does matter, especially to the neighbors who will be looking at your solar ingenuity. When mounting racks are visible, the effect is "industrial." Do you want your home to look like a factory?

The best orientations face due south. As for altitude, the best bet is to orient the panels to the altitude of the sun in the middle of the equinoxes, or around March 20 and September 20. This angle depends on your latitude.

Identifying true south is not as simple as using a compass. Due to imperfections in the composition of the earth, due south rarely matches the compass reading. Here's a simple way to find due south without relying on a compass, or complex "magnetic declination" formulas. Your local newspaper publishes the exact time of dawn and dusk. Calculate the middle of these times; it should be somewhere near noon, but rarely right at noon. Stick a pole in the ground, and at the exact middle time between dawn and dusk, the shadow from the pole lines up with due south.

**Figure 5-9:**
Collector
mounting
options
around your
house.

# Chapter 6

# Calculating Payback on Your Solar Investment

## In This Chapter

▶ Using a system for financially analyzing the solar options

▶ Looking at real-life scenarios you can compare with your decisions

*W*ith every investment, you have options. You don't need to invest in solar energy at all. You can just leave your house the way it is and instead put your riches toward life insurance, stocks and bonds, real estate, and so on. When something breaks, you don't have to replace it with solar equipment; you can just put the same thing back in. But as the financial pros say, you should assemble a diverse mixture of investments to limit your risk, and solar power may be just the ticket to help balance your portfolio — and help the environment in the process.

In this chapter, I give you the big picture on calculating payback — the amount of time it takes to recoup your investment — and help you evaluate the costs, gains, and risks of solar power. I also help you analyze solar investments so you can see what kinds of returns you can expect. Finally, I take you through some real-life scenarios so you can see the numbers in action. (**Note:** Some people prefer to use *return on investment,* ROI, to calculate the value of an investment. ROI is just another way of looking at payback. Because payback is easier for the average layperson to understand , I stick with that form of analysis.) I briefly touch on the paybacks and financial gains of photovoltaic systems in this chapter, but in Chapter 17, I get into a lot more detail on this subject because it's important to those considering installing a full-scale PV system.

Most people are interested in solar power because it helps the environment. Yet the vast majority of solar investments are predicated on achieving a good financial return. In this chapter, I present a way to combine your desire to help the environment with your need for a good return on investment.

# Analyzing Solar Investments

An important factor to consider when deciding how much money to earmark toward a particular investment is the *risk profile.* Some investments entail more risk than others. A bank is much safer than the stock market, for example. When an investment entails more risk, it needs to offer more of a chance for gain to offset the increased risk. Some investors are more risk averse than others, and for these, solar is a better investment.

The following sections outline a system for analyzing solar investments. The goal is to compare different investment options using the same criteria of costs, gains, and risks, and then choose the best one.

## Calculating net costs

The first step in calculating payback is to calculate costs. In this section, I discuss *net costs,* which means the total net amount that you're paying for your solar investment. These days, most solar investments are eligible for rebates and tax advantages, which are subtracted from the "retail" cost of a system. Net cost is simply the starting cost of a system, minus all the subsidies and rebates you can get. It's net cost that you're most concerned with.

### Collecting cost data

The typical solar system has many cost components, and the sum total of all the individual costs must be added up to yield the total cost. Note also that the timing of the costs is important, as well as the timing of the rebates and tax advantages. Timing is important because of the basic idea that a dollar today is worth more than a dollar in one year, the difference being given by current interest rates. So the result of a cost compilation will usually be a timeline rather than a single number. Timing is also important for those who can't come up with the cash to pay for a solar investment all in one fell swoop. It may be the case that you have a few grand available today, and next year you'll get a few grand more, and so on.

In the case of PV, with its high cost, it's especially important to consider how the system must be paid for. Most PV customers finance their system using bank loans and the like. The payments are spread out over time, with monthly costs and up-front service charges. Interest may or may not be tax deductible, and this is an important point to consider.

Consider the following cost factors:

- ✔ **Equipment:** Sometimes equipment costs are spread out over time — for example, if you get financing from a supplier or different parts of a system are delivered at different times. You need to specify the timeline, as well as the dollar value of each outlay.

✔ **Installation costs:** If you're a do-it-yourselfer, you won't have to worry about installation costs. If you contract out, installation costs may be either fixed or charged at an hourly rate, depending on the complexity of the job. In general, be wary of non-fixed types of charges because these can easily get out of hand, particularly with unscrupulous contractors who simply want to maximize the amount of money they make.

✔ **Refuse:** Add in costs if you'll pay to have project trash hauled away.

✔ **Maintenance:** Consider the likelihood that you have to pay for servicing once the system is installed — and when that may occur, so you know whether it's likely to be covered by warranty. Also be aware that when your solar equipment is down, you won't be saving on your monthly utility bills. If it takes two months to get your equipment fixed (rare, but not unheard of), you'll be paying your full monthly utility bills during that two month period.

✔ **Taxes, permits, fees:** Note when such charges are due, as well as the total costs. If you hire a contractor, the contractor usually includes the permit costs in the price. If you're a DIY, you have to pay the costs out of pocket.

✔ **Interest:** If you finance your equipment, the interest is a cost. If you finance with a home equity loan, the interest may be tax deductible.

✔ **Taxes:** Many states have legislation that lets you install solar equipment without paying higher property taxes, but you should find out whether this applies to you (see Chapter 22 for details on state incentives). In rare cases, property taxes may go up as a result of a solar investment.

✔ **Homeowner association fees:** Homeowner's associations may need to review and approve your plans. A charge, sometimes in the hundreds of dollars, is often associated with this. And sometimes the homeowner's association may simply reject your plans altogether. Most contractors won't work directly with homeowner's associations, although they'll help you compile the information that you need when you apply for a permit.

## Subtracting estimated gains and discounts

You may be able to recover some of your expenses right away, so you can subtract that amount from your costs. For example, you may be able to take advantage of rebates and subsidies offered by utilities, state and federal governments, manufacturers, and so on. Or if you're installing equipment for your home business or office, you may be able to depreciate certain items.

You also need to consider both salvage (if you can sell some of the old equipment you'll be taking out) and *appreciation* — the value of your home goes up when you install solar. The appreciation amount depends on the following:

✔ **The cost for a homebuyer to put in new equipment on their own:** Don't expect to get much more than that. However, if the cost of new solar equipment increases, the value of your equipment can also go up.

Equipment can increase in value if inflation becomes significant or if the demand for solar equipment increases (supply versus demand).

✔ **The amount of documentable energy savings achieved with the equipment:** Your energy bills provide proof of the energy savings achieved. Or to be more accurate, your bills reflect how much you're paying for power. The trick here is that a new homeowner may use more or less power than you're using, and their power bills will reflect this. If you're selling your home, prospective buyers will realize that they can afford a larger mortgage since they'll be paying less in utility bills.

✔ **Changes in energy costs:** The more energy costs rise (all energy), the more your equipment — and thus your property — is worth.

✔ **Popularity:** Some things sell because they're the fad. In some real estate markets, a solar home is worth a lot more than the same model with conventional energy simply because people like the idea of solar so much. And the fact that your home is "solar" will attract more buyers, thereby distinguishing your home from others and giving you a leg up on the competition. In some communities, solar and energy efficiency are a big selling point.

If you have a solar home that you want to sell, look for a "green" realtor (not the kind from Mars — the kind with a credential that indicates they have taken special classes that address environmental issues).

Giving a ballpark estimate about how much appreciation you can expect is impossible, but the most likely case is that you'll see your home appreciate around 10 percent more than new equipment costs. Best case, you may double your original investment after five years, if energy rates rise precipitously. Worst case, you may not get anything at all if the equipment is old and obsolete. Old solar water heaters might still be functional, but prospective homebuyers often don't like the" industrial look" of old equipment, plus they realize that they'll be faced with maintenance costs. Twenty years ago, solar water heaters were a big fad, but they were often shoddy and unreliable, and a home with one of these dinosaurs might actually be a liability.

## Figuring out monthly savings

With most solar investments, lowering monthly utility bills is the ultimate goal. Here's how to calculate your potential savings:

1. **Look at your power bills and determine your average monthly energy use (gas or electric).**

   Use the electric bill if the new system will reduce your use of electricity, the gas bill if it'll reduce your gas use.

Suppose, for example, you want to replace an electric water heater with a solar one. You look at your power bills and find that you use an average of 1,000 kWh of electricity per month.

2. **Out of that number, estimate how much energy the system you're replacing uses and how much energy the new system produces.**

   In the example, perhaps an energy audit tells you that water heating makes up 20 percent of your total electric bill (see Chapter 2 for more on energy audits). If you're using 1,000 kWh's per month, then the water heating portion is 200 kWh. With a solar water heater you won't offset the entire water heating bill — maybe only around 75 percent — so you can expect your new water heater to offset around 150 kWh's per month.

   To estimate output of a PV system, use the specifications given for the proposed equipment and modify them up or down depending on your solar exposure (see Chapter 5). Ultimately, you're going to be making an educated guess. The best way to get a good estimate is to ask your professional contractor because he has experience and local empirical data. If you're a do-it-yourselfer, the problem is more difficult. The best way to account for the inescapably hazy nature of projections is to be conservative and see how the economics come out. Then redo the calculations with optimistic numbers and see how the payback calculations turn out. You'll end up somewhere in between.

3. **Multiply your average monthly energy costs by the percentage you found in Step 2 to figure out how much you save per month.**

   At $0.15 per kWh, suppose the total monthly bill is $150:

   $$\$150 \times 20\% = \$30$$

   The result: 75 percent of $30 is $22.50, so the solar water heater therefore saves you that much per month.

If your power company's cost per kWh depends on when you use the energy or how much you use, always use the rate structure that is going to apply *after* you finish the project. For example, if you install a PV system in California, you're likely to get a new power meter and a new rate structure. You may be going from a tiered structure to a TOU structure, in which case you should estimate the cost savings with the TOU structure (see Chapter 17 on how to calculate your monthly savings with a PV system in the different rate structures). Rate structures are changing all the time, but it's likely that most areas are heading for the TOU structure that California currently uses. It's also true in California that utilities have to pay you the same rate they charge you, which isn't necessarily true in other parts of the country. The devil is in the details.

As energy costs rise, your savings increase even more. The Energy Information Administration reported that from 1996 to 2007, electricity energy costs increased an average of 2.2 percent per year (www.eia.doe.gov/emeu/steo/pub/contents.html), and the change may be even greater in the future — or in your region of the country. In California, electricity rates have gone up an average of 7.6 percent per year over the last decade, and it's likely that rates will rise even more than this when government gets truly aggressive about global warming and pollution in general. At any rate, it's safe to say that 2.2 percent average yearly increase is a thing of the past (just like incandescent light bulbs).

## Putting the numbers together: Figuring payback

*Payback* is the amount of time you need to hold on to your investment for it to pay for itself. In the case of solar, almost the entire investment is upfront, meaning you have to invest cash before you see a single dime of savings.

Payback is measured in years or months. You simply take the net costs of your solar system (costs minus discounts and appreciation — see the earlier "Calculating net costs" section) and divide it by your anticipated monthly savings (see "Figuring out monthly savings"). For example, if you invest $2,000 in a solar water heater that saves $30 a month in electric costs, here's your payback period:

$2,000 ÷ $30/month ≈ 67 months, or 5.6 years

This is a simple example and ignores not only the time value of money, but also the fact that your savings are going to increase with each and every increase in utility rates. With any kind of solar system, you save more and more each year, but it's impossible to predict exactly how much (unless you have a solar powered crystal ball to consult).

---

## $1 today versus $1 tomorrow

Strictly speaking, when doing payback calculations, the time cost of money must be taken into account, which basically means that you discount the value today of a receipt of a dollar in the future.

Suppose that you have a choice between two alternatives: $1 today or $1 a year from now. You take the $1 today, of course. But how much would you be willing to trade that dollar today for a payment of $1 in one year? The answer is given by the current interest rate, or even more precisely, by what the interest rate is going to be over the course of the next year. Say that the interest rate is 6 percent. Then you would trade a dollar today for $1.06 a year from today.

In calculating payback, it works the other way around. For example, if you're going to save $100 a year from now, that savings is only worth $94 today (at 6 percent interest).

In the following calculations, I ignore the time cost of money, for simplicity sake. A way to account for the time cost of money, without making the computations overly complex, is just be very conservative. The net effect is the same.

# Making online calculations

Fortunately, a number of resources can help you gather numbers for the calculations, and some even walk you through an automated calculation. You just answer their questions, and they crunch the numbers:

✔ **Database of State Incentives for Renewable Energy:** Visit www.dsire usa.org. DSIRE is a comprehensive source of information on state, local, utility, and federal incentives that promote renewable energy and energy efficiency.

✔ **Clean Power Estimator:** Check out www.consumerenergycenter.org/renewables/estimator. Clean Power Estimator is an economic evaluation software program the California Energy Commission is licensing for use from Clean Power Research. The program provides California residential and commercial electric customers a personalized estimate of the costs and benefits of investing in a photovoltaic (PV) solar or small wind electric generation system.

✔ **The Energy Grid:** See www.pvwatts.org. This tool allows you to estimate the number of watts of power you can expect from your solar system in a given geographical area.

✔ **Find Solar:** Go to www.findsolar.com. These estimators are driven by extensive databases of solar ratings, regional weather conditions, and applicable incentives. Very practical.

## Analyzing risk

No investment is risk-free, and some investments entail a lot more risk than others. The following elements can affect the returns you get on your investment:

✔ **Repairs and maintenance:** Solar equipment is not simple when it's doing a big job. And being outdoors all the time is wearing, especially in extreme climates (how'd you like to be strapped to a roof 24/7?) You can't anticipate when something will break down, and if your equipment has problems after the warranty expires, your costs go up. (You can buy extended warranties, but they cost a lot and aren't usually worth it.)

✔ **Efficiency decreases over time, just as it does for autos and humans and pretty much everything.** PV panels may see a 10-percent decline in production after 10 years of operation. It's harder to estimate the decline in solar water systems, but it's on the same order, unless you're using hard water or have well water with a lot of sediment, in which case you may see a 30 percent decline in 10 years. In general, use 10 percent per 10 years in your calculations for a good approximation.

✔ **Lifetime:** Eventually, every system will simply not be worth running anymore. PV systems are warranted up to 25 years, water systems for up to 15. However, it's not as bad as that, because what the warranties really cover is guaranteed performance. You will probably be able to run your systems for much longer than the warranties, but they simply won't be putting out the same performance. The 25 year warranty for PV panels, for example, says that they'll output 80 percent of their original power level after 25 years. In practical terms, this is a tough number to verify. Plus you probably won't be living in the same house in 25 years. But it's reassuring to know that a PV system is built to last, and most of them do.

✔ **Inexperience:** If you design and install a system yourself, the performance may not be as good as you had planned. The experts know all the little tricks, but you don't. Savings can suffer, and you may have to pay an expert to come in and set things straight.

✔ **Newer, more efficient technologies:** People like new things because they're new. The bottom line with a solar system is the performance it's putting out, but it may be true that a homebuyer simply doesn't like a system that's 10 years old simply because it's 10 years old. Newer equipment is always more efficient, plus it's shinier and may be better looking. This increased efficiency rarely results in a situation that merits tossing out old equipment that is still working, but it may make a difference to a prospective buyer.

## Accounting for the intangibles

Good decisions aren't just predicated on dollars and cents. You also need to consider the intangibles, such as aesthetic beauty and pollution mitigation, among others. Not everything comes with a price tag, but you certainly need to factor in how much these elements are worth to you. You can do this by either assigning dollar values, as best you can, or making a list of priorities so that if two potential options have the same numerical score (payback calculation), one option will win over the other based on the intangibles.

### Allowing for drawbacks

You should account for any inconveniences that a solar system will entail. Note whether your system is going to require more work on your part. Solar water heaters, for example, take maintenance, and you need to monitor their performance on a periodic basis. You may need to evacuate the system in extreme weather to prevent freezing. Decide whether you're going to do it or hire someone else to do if for you. In addition, when you install a solar system, you're introducing dangers into your house: electrical shocks. Scalding water. Falling off a roof. Discovering stepladder malfunctions. Think about whether you're okay accepting these risks.

Also factor in your feelings about the look of your house and how that may affect resale value. Some solar equipment can be ugly. Do the pipes running up the side of your house look industrial? Do photovoltaic panels make your house look unearthly? On the other hand, some solar equipment, like integrated PV panels that match your roof, can actually make your home more attractive.

### Considering priceless benefits

Think about visual improvements. For example, if you invest in a maple tree, you get shading in the summer, which is a form of solar system performance, but the greater pleasure is having a nice tree near your house. When you invest in window coverings, you enjoy better insulation by virtue of the fact that you can shade your home's interior from solar radiation, but the greater gain is probably in the remodel, and the fact that your home is more attractive and congenial.

Don't forget environmental improvements: When you invest in solar, you're saving a lot of carbon dioxide, and you're also helping with national energy independence. Some people, for whom this is very important, install solar equipment whether or not it makes financial sense. This is referred to as the _green premium_.

# Examining Real-Life Scenarios

Compare your own solar investment decisions with these examples and you can get a good idea of how your investment stacks up. In most cases, just substitute in your own numbers. In some cases, you may have to adjust the model to fit your exact situation. (Go to Chapter 17 to get a lot more detail on how to calculate the payback for a photovoltaic (PV) investment.)

Decisions often come down to the details of your rate structure. You can find it on your power bill. You may be able to change your rate structure with a simple call to the utility. Some allow for choices. In fact, you may be able to save on your monthly power bills without using less power or investing in solar or any other energy efficiency equipment. How? Simply by changing your rate structure. This may sound like a sleight of hand (and it is), but it's a real effect.

_Note:_ For payback calculations for full-scale PV systems, head to Part IV, which is devoted to that topic.

## Supplementing an existing water heater with solar

To calculate payback, first find the monthly energy savings. Suppose Household X uses an average of 1,000 kWh of electricity per month. At $0.15

per kWh, the monthly cost is $150. An energy audit determines that water heating comprises 18 percent of the total electric bill, at a cost of $27 per month (see the earlier "Figuring out monthly savings" section for details on such calculations). With the solar water heater, that amount drops to $7, so $20 is its monthly savings.

Then determine the net cost of the equipment. A solar water heating system costs $2,000, including parts and installation. A federal government tax credit of 30 percent is available, for a total of $600. The net cost is $1,400. The warranty is for 5 years, so the homeowners won't have to pay maintenance costs for the first 60 months.

To recover the initial investment of $1,400, Household X will need

$$\$1,400 \div \$20/\text{month} \approx 70 \text{ months}$$

That's a good investment (and keep in mind that this calculation doesn't include annual utility rate increases). A solar water heater may last about 20 years, and Household X is making a profit within the first 6 years.

Here's another factor to consider. If you're displacing a gas or propane powered water heater, prices may fluctuate wildly from year (25 percent is not unheard of, as anybody who pays for these commodities can attest). In this case, it's even more speculative to calculate cost savings. But in this case, the hedging factor comes into play even more because you're reducing your exposure to the wild fluctuations in price.

*Hedging* is making investments that reduce risk. If you invest in solar, you shield your utility bill from energy price fluctuations. If you put a solar system on your roof that reduces your energy bills to zero, that's exactly where they'll stay, even if the cost of energy quadruples. That's a powerful form of hedging. It allows you to smooth out the fluctuations in your monthly utility bills, which helps you keep your budget in line with expectations (especially attractive for people on fixed incomes).

## Putting money in a bank or stock market

Here's how some other investments stack up: Put $1,000 in a savings account at 6 percent, and you gain compounded interest. You have $1,000 in interest *12 years* after you make your original investment, which is the payback period. At any point in time, you can pull your money back out of the bank, with zero risk.

## Accounting for annual rate increases

To account for annual rate increases, redo the calculations using the projected increases. In this example, assume that projections show the rates increasing 10 percent per year. (I chose this number simply to make the math easy, but 10 percent annual increases aren't out of the realm of possibility. In fact, some people believe that even this seemingly high number is going to be low.) Here's how the annual rate savings stack up now for the system outlined in the section "Supplementing an existing water heater with solar":

| Year 1 | $7 per month | $84 per year |
| Year 2 | $7.70 per month | $92.40 per year |
| Year 3 | $8.47 per month | $101.64 per year |
| Year 4 | $9.32 per month | $111.80 per year |

...and so on. Now the payback for that system goes down to around 58 months, which is a much better investment and reflects the more likely scenario.

If you put your money into the stock market and get a return of 12 percent, the payback period is only 6 years. But the value of your stock could also go down in value, maybe all the way to zero. Because the risk is much higher in the stock market, you insist on a higher potential gain or else you'll just put your money into a bank.

A good general rule is the *Rule of 72:* To find the number of years required to double your money at a given interest rate, just divide the interest rate into 72. For a 12-percent return:

$72 ÷ 12 = 6 years

For 3 percent return, the payback is 12 years.

## *Accounting for pollution*

Suppose you're installing a solar water heater in an area where the climate is northern. There isn't as much sunshine, and the winters are colder. Cost savings are $13.50 per month. For this example, assume constant energy costs, which is a simplification. (The point here is simply to illustrate how you can account for pollution, so we don't need to nitpick the details.) When the monthly savings are cut in half, the payback period for a $1,400 water heater is about 104 months. This isn't much different from the return (or payback) you can get at a bank, where there's zero risk and you can take your money back out whenever you want. When you invest in solar, you're stuck with what you've got. This is a bad investment and is typical of the price you pay when you don't get a lot of sunshine.

But how much is it worth to reduce pollution? Ten dollars a month? Twenty? How much would you pay to completely eliminate your carbon footprint altogether? $1,000 a year?

This may sound like a theoretical question, but it gets to the heart of the issue: How much are we willing to pay to reduce pollution? The fact is, renewable forms of energy have a very difficult time competing against fossil fuels. In general, the cost of alternative energy is higher than the fossil fuel mainstays. When the government subsidizes alternative energy systems, we all pay in the form of higher taxes (or higher deficits, since the government doesn't like to raise taxes, even though they like to spend more money). So the question is not rhetorical, or theoretical; it's a real question.

Out of your total carbon footprint (see Chapter 2), figure out how much $CO_2$ the old water heater contributes. The typical carbon footprint is around 40,000 pounds per year. The typical footprint from a water heater is 3,000 pounds per year, or 7.5 percent of the total (this total includes not only the pollution you generate at your home, but your transportation as well).

If you're willing to spend $1,000 a year on pollution mitigation, you'd spend 7.5 percent of this on heating water. That's $75 per year, or $6.25 per month. Add this to the savings column, and instead of only $13.50 per month, the real savings is $19.75. This makes the payback 71 months, which is a much better proposition:

$$\$1,400 \div (\$13.50 + \$6.25) \approx 71 \text{ months}$$

If you'd spend $2,000 a year eliminating pollution, total cost savings are $27 per month, which is all the way back to the original example with a payback of 52 months (see the preceding section). This is a good investment.

So how much are you willing to contribute?

## Reaping rewards of rising energy costs

What if energy costs rise 12 percent per year instead of staying flat? That percentage is higher than historical norms, but it's not difficult to imagine how this might come to be, with all the pressure coming to bear on the energy markets. When energy costs rise, each year your savings grow. Perhaps your $1,400 solar water heater saves you $27 per month the first year. But the next year the savings grow to $30.24 per month ($27 × 1.12). Then $33.87 ($30.24 × 1.12), then $37.93, then $42.49, and so on. Look at the yearly savings to see how they add up:

| Year | Monthly Savings | Annual Savings | Cost Left to Recoup |
|------|-----------------|----------------|---------------------|
| 1 | $27 | $324 | $1,076 |
| 2 | $30.24 | $362.88 | $713.12 |
| 3 | $33.87 | $406.43 | $306.69 |
| 4 | $37.93 | $455.20 | −$148.51 |

The payback in this case is 44 months (down from 52 months if the energy costs were to stay the same).

Regardless of how high energy costs go, the amount that's being spent on heating water is locked in. This is a form of *hedging,* or making an investment that reduces risk.

Here's another benefit, not quite as tangible but still very real: There's no reason to take shorter showers or skimp on that bath, no matter how high energy costs may go! A solar heating system has so much capacity; not using all of it saves nothing. As to whether you want to save water, that's a different story.

## Calculating for different rate structures

In calculating how much you'll save on your monthly power bills when you install a solar system, your rate structure is of critical importance. In this section I describe the general dynamics of the different rate structures that you may be working with. In Chapter 17, I present examples that pertain specifically to PV systems.

### Thinking about tiered rate structures

In a *tiered rate structure,* the more energy you use, the more you pay per kWh. So in California, for example, not all watts are created equal. Some tiered rate structures are very punitive, with the highest rates three or four times the base rate. Solar system savings subtract from the highest tier first. In such a rate structure, cost savings can easily be twice as much. Payback is therefore half as long. Here's a typical tiered rate structure in Northern California:

Tier 1, the first 1,033 kWh of consumption costs $0.1153 per kWh.

Tier 2, the next 310 kWh of consumption costs $0.13109 per kWh.

Tier 3, the next 723 kWh costs $0.25974 per kWh.

Tier 4, the next 1,033 kWh costs $0.37866 per kWh.

Tier 5, which you don't want to be paying, costs $0.44098 per kWh.

Note that the highest tier is around four times more expensive than the lowest tier. To make matters worse, the upper tiers are generally raised more, in percentage terms, than the lower tiers (this is like the progressive income tax rates that we all love so much).

A solar water heater may offset 14 percent of the household energy use, but this may comprise more like 35 percent of the total dollar value in a typical tiered rate structure. It all depends on how much power you use per month.

In a tiered rate structure, small investments work best. As the size of your investment grows, the payback gets worse because you get less and less return on your investment simply because you're offsetting the lower tiers with a larger system. However, for people who anticipate large rate increases, regardless of the tier, this effect is reduced.

One more point to consider: In a tiered rate structure, your highest and lowest bills fluctuate to a larger extent. In the mild weather months, like May and October, you use less power and you pay for less upper-tier energy. Even though you may only use 75 percent as much energy in May as in July, your bill may be twice as much in July. When you install solar, you shield yourself from this effect.

### Considering Time-of-Use (TOU) rate structures

In a time of use rate structure, you're billed for power based on the time of day that you use the power. For example, from 12 noon to 6 p.m. (peak time), power may cost three times as much as power used in off-peak hours. In general, weekends and holidays are off-peak. The basic goal with TOU rate structures is to dissuade power customers from using power during peak times, and the effects are profound. (For more on why peak rate structures are used and becoming more common, an interesting subject that may become very important in the years to come, read my book *Alternative Energy For Dummies,* Wiley.) If you're in a TOU rate structure, you may not even be aware of it. Check your power bill.

Accounting for payback in a TOU rate structure is very difficult because you have to make some assumptions about when you use your power. For instance, if peak time is noon to 6 p.m. what percentage of your total power consumption occurs during that time? Decide what the percentages are; then use the different rates to determine the total costs during peak and off-peak. Be conservative in your estimates.

Some power usage is unavoidable: refrigerators need to run continuously, for example. But you can still probably decrease your monthly utility bill by reducing your power consumption during peak hours. Avoid taking hot showers during peak hours. Use your air-conditioner less (cool your house way

down in the mor`...` turn off the AC at noon; while the temperature will s`...` find that it's tolerable). Avoid using
pool pumps`...`

PV syst`...` use they output the vast
majo`...` minimize your use of
po`...` cranking away full tilt,
you c`...` able degree. Head
to Chapt`...` k in a TOU rate
structure fo`...`

## Appreciating `...` increase in valu`...`

Suppose a homeowner decides to se`...` s after installing a
solar water heater. Energy costs are risin`...` savings from the solar
water heating system are now $48 per month, `...` per year. A homebuyer
will pay more for the home because of this built-`...` cost reduction.

How much more? Let's say new systems cost $2,800, and tax rebates are a
thing of the past because everybody and their brother are now in the market
for solar. A lot of work is involved, and most buyers don't want to do it.
They'd have to read some highly technical books, for example (unfortunately,
they didn't know about this handy little guide).

Even more importantly, most homebuyers use mortgages, where balancing
monthly payments with a fixed income is the game. Forty-eight dollars per
month in cost-savings translate into $48 that can be spent somewhere else. A
buyer could get a larger mortgage, for example. For $50 a month, after taxes,
you can borrow $14,000. So although putting in their own solar system may be
an option (after the purchase of the home) for homebuyers, they also have a
strong incentive to purchase existing solar equipment with their mortgage.

In terms of appreciation, the homeowners are likely to get about 125 percent
of the price of new equipment. For the example, a 25-percent appreciation
would be $3,500 (that is, $2,800 times 125 percent). Suppose that with rising
energy costs, the original investment of $1,400 paid for itself in 44 months
(see the preceding section). After that, the cost savings were all pure profit.
At 60 months, the sellers get a further profit of $2,100 at the sale of the house.

To find out more about how a home appreciates with a solar system, consult a
realtor in your area who can give you an idea of how much more solar homes
are selling for than conventional homes.

## Financing solar investments with a home equity loan

It's common to save more on energy costs than the payments on a mortgage equity loan used to purchase a solar system. You pay no upfront cost when you finance an investment. You do have to pay off the loan at some point, but from day one, you save more than you pay. Your cash flow is in the black.

Suppose you borrow $1,400 for a solar water heater with a promise of $27 per month in savings. An interest-only home equity line of credit at 6-percent tax-deductible interest for the $1,400 is only $8 per month (from mortgage payment tables). That's a net gain of $19 per month. If the interest on that equity loan is tax deductible, the savings are even more.

Here's the payback period for this scenario:

$$\$1,400 \div \$19/\text{month} = 74 \text{ months}$$

If the interest is tax deductible, the payback may be as low as 40 months, depending on your tax bracket.

At some point the loan is repaid (principal payments usually kick in after five to ten years), although most equity loans get paid off when the house is sold or refinanced. When it's sold, you often get more for your solar equipment than what you paid (see the preceding section for details).

## Replacing broken water heaters

When your existing water heater crashes, doing nothing is not an option. Now you must spend money, probably a good chunk. How much more do you have to spend for a solar system? Probably not much. Now's really the time to go for it.

For example, a solar water heating system costs $3,000, and a conventional water heater costs $1,400. The difference is only $1,600, whereas if you take out a working water heater in order to upgrade to solar, the real investment is the entire $3,000. (And if you factor in rebates and subsidies, the difference may be zero.)

If you have old equipment, you've probably paid to have it fixed and you'll probably have to do so again in the future. If you buy a new solar system, you won't have to worry about it breaking down at the worst possible time (which is when things usually break, as per Murphy's Law). This is a form of psychological hedging called peace of mind.

# Diving in to swimming pool solar heaters

A solar pool heater does *not* save money — except when you compare it to all the other options for heating your pool. Solar collectors load the filter pump (make it run harder) so it costs more to run. Sometimes a lot more. If you want a lot of heat, you have to run the filter pump for upward of eight hours a day. This extra pump time is solar-system cost, and if you're in a tiered rate structure, it can add up to quite a bit. If you're in a TOU structure, you also get dinged because you need to run your pump at peak times (midafternoon, when it's sunniest).

When you invest in a solar pool heater, you pay thousands to get higher power bills. It's not worth trying to do a payback analysis on a deal like that. The only gain is comfort, and that's hard to value. But people pay a lot for solar pool heaters, and in some areas you see them on literally every single home that has a pool. Ask an owner, and he or she will probably tell you they're great. Pools cost upward of $25,000. Another $3,000 to make it much more enjoyable isn't a bad investment. Your swimming season will be extended — a solar heater may triple the amount of time swimmers actually spend in the pool, as opposed to shivering around it.

The biggest plus with solar pool heaters is that other options are insane: a big, gas heater that gobbles propane, or electric, with a 24/7 spinning power meter. (You could light a big fire and boil water in a huge pot, but that's labor intensive.) A solar pool heater not only heats your pool much cheaper than the other options, but the pollution effects are infinitely better. In terms of payback, the only benefit of a solar poor heater is that it makes your original investment in the pool look much better, and that counts for a lot, in some people's view.

# Blowing hot air

The investment: an attic vent fan powered by PV panels. Equipment costs $350, and installation takes two hours and requires brains, ladders, roof climbing, and decent tools.

An energy audit determines that the fan will save $25 per month in air-conditioning costs over a four-month summer period (see Chapter 2 for details on energy audits). At $100 per year, the payback is only 3.5 years.

You also boost your comfort. The house is a lot more comfortable on the days when it isn't hot enough to turn the AC on. And because the fan lets hot air escape, you get more of these days.

The alternatives to installing a solar fan? Not installing anything at all and paying for all that extra air conditioning, or installing a hard-wired fan for more than $1,500 because you need a licensed electrician to do the installation, as well as county inspections and permits.

In hot climates, a solar-powered attic vent fan can be one of the best investments you ever make. And with the kits available today, nearly anybody can install a simple version in only a few hours or less.

## Working with solar in your home office

You can go off the power grid with a home office. A remote cabin solar power kit costs $1,000, with 12 volts DC (direct current) output and a battery so power is always available. You can use the system to run DC lights, a vent fan, coffee maker, radio, and a PC laptop computer with printer. A small inverter provides minimal 120 volts AC (alternating current) power for telephones and other support gear.

The total investment, including wiring and DC equipment (lights, fans, and an AC voltage adaptor), is $1,500. Installation is extra, but with the kits available, a do-it-yourselfer can tackle the job. And because the solar equipment is used for a business, it's depreciated, which reduces taxes.

The system saves around $30 per month in utility-provided electricity, so the payback is 50 months. When all taxes are taken into account, the payback can be as low as 30 months.

# Chapter 7

# Installing Your Solar System

*In This Chapter*

▶ Choosing your best options

▶ Getting the most value from the projects you choose

▶ Keeping an eye on safety

*Y*ou may be chomping at the bit to get moving, but you want to make sure that you start out in the right direction. In this chapter, I present a number of ways to gauge your decision. I also tell you what to do after you pick a project so you can make the right installation and safety decisions.

## *Paying Attention to Your Pocketbook*

Solar projects can cost thousands of dollars, but you can find ways to harness sun power on almost any budget. Chapter 6 can help you figure out costs in the long run, and Part IV of this book explains some funding options for the larger projects. Here, I lay out initial costs. Table 7-1 gives you an idea of what you can do, given your budget.

| Table 7-1 | Projects and Budget Considerations | |
|---|---|---|
| *Initial Cost* | *Project* | *See Chapter* |
| Free+ | Plant a tree in your yard | 8 |
| $8+ | Toys and gimmicks | 9 |
| $10+ | Solar lights for your yard | 8 |
| $10+ | Sunscreens for your windows | 9 |
| $18 | Portable solar shower | 10 |
| $30 | Water purifier | 10 |

*(continued)*

## Table 7-1 *(continued)*

| Initial Cost | Project | See Chapter |
|---|---|---|
| $30+ | Blinds, drapes, ceiling fans, sunscreens | 9 |
| $50 | Solar cooking | 9 |
| $60+ | Off-grid reading lamp | 9 |
| $80 to thousands | Solar fountains | 8 |
| $80+ | Radiant barriers in your attic | 9 |
| $100+ | Home security system with great comedic potential | 9 |
| $220 | Solar light tube | 9 |
| $320 | Attic fan, do-it-yourself | 13 |
| $600+ | Attic fan, installed | 13 |
| $600+ | Automatic awning for a patio, porch, or picture window | 9 |
| $1,000+ | Off-grid office, barn, workout room, and the like (without HVAC) | 18 |
| $1,700 | Domestic hot water preheater, do-it-yourself | 12 |
| $2,000 | Swimming pool solar heater, do-it-yourself | 11 |
| $2,000 | Hot tub solar heating system | 11 |
| $4,000 + | Fully installed swimming pool heating system | 11 |
| $4,000 | All climate, professionally installed solar hot water heater | 12 |
| $7,000+ | Solar-powered radiant heat floor system | 12 |
| $15,000+ | Intertie PV system | 16 through 19 |
| $150,000 into the millions | Solar home | 21 |

# *The Best Projects for Do-It-Yourselfers*

For those who like to work with tools, the following projects provide the best opportunity to hammer, screw, cut, and curse.

## Reaping the rewards of photosynthesis

Start a vegetable garden and use solar energy to feed your family. You want a good start? Put a tomato plant into a big wooden barrel and give it decent food and water. You even can get a small automatic valve so you don't have to do the watering. Certain varieties of tomatoes are relatively forgiving, and they provide even if you're lax. But if you're diligent, you can get tons of fat tomatoes. Thousands of different varieties are available. Get a rugged, proven performer; forget fancy unless you're a good gardener. (If you're really interested in growing things, check out *Gardening Basics For Dummies,* by Steve Frowine and the National Gardening Association.)

Full-scale gardens are dubious in terms of payback. They're expensive and a lot of work. But there's nothing like the feeling you get when the entire, incredibly delicious dinner came from the garden — even the rabbit stew.

Solar cooking is a lot of fun, and nearly anybody can do a good job of building an oven (see Chapter 9). It's rewarding, and the food tastes good. Ovens are cheap enough that you can make mistakes and rebuild without being out much hard-earned cash. They're simple enough so that most do-it-yourselfers can develop their own designs after they learn the ropes. Designing your own equipment is especially fun when the final product is getting better and better with each effort. You can build a solar oven for less than $40 and with very few tools.

The ways you can create a solar fountain are diverse; plus you get to work with all kinds of interesting physics. In this project, you can make something truly beautiful out of next to nothing. You can even use the technology to provide a domestic water supply for a remote cabin. (See Chapter 8 for details.)

Another good option is an attic vent fan; it makes so much sense for hot climates, and it has good payback if you do it right. (Flip to Chapter 9.)

Solar water heaters can be ideal financial investments, especially if you're in a tiered rate structure (see Chapter 6), and many can be installed by the do-it-yourselfer. You can typically save 40 percent of the cost of a system when you do your own installation. You can install a complete system for under $1,500 in equipment. Right now, the feds give you a 30-percent tax break, so you can get in for about a grand. (See Chapters 10 and 12.)

# Considering the Best Overall Investments

You want the most return for your investment. Or maybe you're looking for the quickest payback for the least amount of effort and cost. These projects are the winners in this category.

- ✔ **Conservation:** Perform an energy audit and perform all the energy conservation measures that your audit suggests. Energy conservation saves money, reduces pollution, and helps make going solar much more practical. See Chapter 3.

- ✔ **Interior blinds and window coverings:** These items yield the best payback because they have so much influence on your environment. The payback's not strictly monetary, but who says it has to be? (See Chapter 9.)

- ✔ **Solar screens:** If your climate is hot and sunny, solar screens (Chapter 9) are the easiest, biggest payback for the smallest investment. For a few dollars a year, you can keep a lot of heat out of your house, save on energy costs, and make a room more comfortable on sunny days. Of course, the overall value depends on the installation; yours may be overly difficult or impossible due to window configurations or accessibility. While large, southern-exposed picture windows are the best candidates because they have the greatest affect on solar radiation, they're sometimes difficult to completely cover without creating some ugliness in the process.

- ✔ **Landscaping:** Plant a tree or two. A few years later, you'll understand what a real payback is. Not only will you reap the benefit of a beautiful yard, but the cooling effect can also be dramatic. I have a personal bias here, I should admit; I favor beauty over financial return, and I encourage you to as well. After all, beauty is lasting, and it shares with everybody equally. Flip to Chapter 8.

- ✔ **Solar battery chargers:** They're easy to use and inexpensive, and when you charge your own batteries instead of purchasing disposables, you definitely save a lot of money. See Chapter 9.

- ✔ **Solar attic fans:** Solar attic fans (Chapter 9) are great for the experienced do-it-yourselfer, especially if you live in a hot climate.

- ✔ **Intertie systems:** Intertie system economics depend a great deal on subsidies and rebates and tax breaks, but when they're advantageous, you get a lot of value out of a single, overriding investment. You get guaranteed power rates (free) forever, and this locked-in rate may be very, very important. But you also save a lot of carbon dioxide pollution, which should be a factor in your decision (see Chapter 20).

- ✔ **Hot water heaters:** They can get a big return in tiered rate structures, and they can get really big returns when you do your own installation. Chapters 10 and 12 can fill you in.

✔ **Solar light tubes (tubular skylights):** Solar light tubes (Chapter 9) are a good do-it-yourself project with a potentially big return. When you put some thought into where one should go, the effect can be very dramatic for the cost. Try to find somebody who already has one who can show you how it works, because they have a definite character.

# Getting the Most Out of Your Equipment

Here's my best advice in the entire book: I call it the measure-twice-cut-once rule. When working a project, do one part at a time, and then put it aside for one week. You can do this sort of thing with any project, and I guarantee it'll come out better. The more of a novice you are with the particular skills of a project, the more this rule applies.

For example, suppose that you want to put a PV system up on your roof. Phase 1: Decide on all the options, collect all the financial numbers, then put it aside and forget about it. A week later, rethink the whole thing. Then calculate all the paybacks and scenarios. Make a plan. Forget it again for another week. Then start making calls to contractors. Talk to owners about costs and performances. Get a feel for the supply market — who's good, who's not. Wait a week. Make a decision; contractor, money, where it's coming from, and the like. Wait another week. Cast the decision in concrete and start moving.

Here are some other tips:

✔ **Buy kits when you can.** Even when a contractor does an installation, ask for complete kits whenever possible. System performance is better, and you can rely on the instructions and support. And you don't run into any arguments about whose parts aren't working right when a cobbled-together system is performing poorly. Contractors generally favor certain manufacturers and types of systems, so you may want to use this tip in selecting a contractor. If two contractors are tied, all things being equal, favor the one who plans on using a kit that's withstood the test of time.

✔ **Get a price for installation, even if you want to do it yourself.** Sometimes the price difference between buying a kit and installing it yourself isn't much different from having the entire thing professionally designed and installed. Professionals can get parts more cheaply than you can, and they can install quickly and efficiently. (On the other hand, what's the point of calling yourself a do-it-yourselfer if you don't?)

✔ **Get more than one bid, preferably three.** Solar is growing by leaps and bounds. Things are changing fast, and options are expanding. When you get three bids, they'll probably be different — maybe a lot. Understand why. *Growing, inventive markets are always buyer-beware.*

> ✔ **Don't automatically go for the lowest bid — it may be the highest bid in disguise.** A lot of new characters are getting into solar energy these days. Many are great. The industry is attracting new blood. On the other hand, some newcomers are setting up shop only because so much work is available. Good contractors always have references who'll generally be happy to put in an encouraging word. If your contractor is vague about references, proceed with caution. Newcomers with slim references may be the best option, but there's more risk, so you should expect their price to reflect it.
>
> ✔ **Get references.** Try to find somebody who already has the system you're thinking about purchasing. What are his recommendations?

# Playing It Safe

Honestly assess your skill set in terms of handling dangerous situations. If you're not a great do-it-yourselfer but you like trying and know you'll get better, you can do a lot of things where nobody can get hurt. Overall, when it comes to safety, don't mess around. This section outlines some situations to avoid, things you shouldn't do unless you really know what you're doing, and areas where you should be particularly cautious.

## Intertie PV-generating systems

Only people who really, truly know what they're doing in all aspects of the job should install a grid intertie system. In such a system, your homegrown power generator is connected into the public utility grid so you can sell back excess power. You can't interface with the utility company unless you're highly qualified — and licensed. The utility company also has a lot more lawyers than you do, and it will sue you in a heartbeat if you do anything slightly out of its realm of comfort.

Don't wire up or work on a PV intertie system if you're not qualified: The voltages are high enough to kill.

## Plumbing and electrical work

Avoid plumbing or electrical work if you don't understand the code requirements or safety hazards. For example, any power system that puts out 120 volts AC is potentially dangerous. If you don't understand high voltages and how you can get a shock, don't mess around with this stuff. At all.

Bad installations can create dangers that don't have to be. Books can take you only so far into the project world. Real-world experience counts for a lot. This doesn't necessarily mean you shouldn't try something you've never done. You can get good explanations for how to do things from a clerk in a hardware store, and you can do many jobs on that basis. Just understand that patience and consideration are at premium when you're forging new territory.

If you perform work that falls under county code, you may have problems selling your house someday if an inspector determines that you haven't met the standards. If you do an installation incorrectly, you may have problems with your homeowners' insurance. Code exists for a good reason: It's not just so the government can give you a hard time — it's to keep things safe and consistent. You wouldn't want to buy a house with sub-code work, and you don't have to. A seller must bring it up to code. If you sell your house with the promise of solar energy, you'll be expected to turn over equipment that's up to code and does what it's supposed to.

You can plumb all you want on a system that's not connected to your domestic supply, such as an outdoor fountain or a swimming pool. But when you install a solar water heater, you may want to have a plumber install the pipes at the very least (see the upcoming "Water heaters" section). You can buy all the equipment, mount it into place, and then get a plumber to connect the pipes.

Also, many of the governmental/tax incentives require a system to be installed by professionals with certain minimum certification credentials.

## Solar panels

Climbing around on roofs is statistically the most dangerous part of any installation. Injuries can be severe, maybe even fatal. Do you belong up on a roof? Not just climbing around, but pulling heavy weights and tools and going back and forth a lot? If you have a flimsy ladder, you're asking for big trouble. After all, solar panels are heavy and awkward. (Imagine hoisting a heavy piece of plywood around in the wind while you're balancing on a roof, and you get the idea.)

Solar water heating panels can be particularly heavy when full of water. If you don't know your roof can take the weight, either find out from an engineer, which costs you money, or forget about it. Imagine what a stinking mess a caved-in roof would be, especially with hot water dripping all over! You probably wouldn't be a lifelong solar fan.

Also pay attention to local ordinances. Sometimes building codes forbid solar panels in view from the street, although more and more state legislatures are banning this limitation.

## Water heaters

Don't install a solar water heater if you don't understand its engineering, particularly its dangers and maintenance requirements.

Water heaters can be problematic in hot weather. When a solar panel sits in the direct sun long enough, the water can sometimes rise to over the boiling point. Solar panels have been known to rupture from steam pressure, which can burn (not to mention cost a lot to fix). Or the pipes can fill with superheated water and burst inside your house.

Always keep children away from solar water heating systems and the associated pipes. Also, add some safety measures to your valves. Most solar water heaters have valves for controlling flows and evacuating equipment for various reasons, such as performing maintenance or keeping water from freezing in the pipes in cold weather. Children love to play with valves, so buy valves with padlock loops to prevent children from reaching up and twisting a valve with superheated water.

## Batteries

Batteries are safe, effective, and reliable when used properly, but they can also cause worse injuries than electrical shocks. If the terminals of a big battery get shorted, a tremendous arc of current flashes, precisely the same as an arc welder. You don't want to be around when it happens.

Unsealed batteries can emit noxious fumes. They can corrode, so take special precautions when you dispose of them. In order to wire up a safe and efficient battery circuit, you need to know exactly what you're doing.

Buy a system with batteries in complete form — all these battery dangers should be adequately addressed.

# Part III
# Applications Aplenty: Projects from Small to Large

The 5th Wave    By Rich Tennant

I love how you've handled the lighting in here.

## In this part . . .

**H**ere's where the fun begins. This part showcases real, hands-on solar projects. You can start in your yard or patio. Use a lot of fun little gadgets and gizmos around your house or give them to your kids for an educational play session.

Controlling the sunlight in your house is easy and inexpensive, and it can have a major impact. You can heat water with the sun any number of ways to help around your house. When you understand your home's ventilation schemes, you can install solar fans and make your house much more comfortable, as well as keep the monthly power bill down. And you can buy solar systems that can power your boat, RV, or remote cabin. I touch briefly on wind and water resources, the neglected cousins in the solar power family. And finally, one of the most influential projects you can do for your home is adding either a sunroom or a greenhouse. All of which are covered here.

# Chapter 8

# Digging In to Landscape Projects

*T*he great outdoors is a good place to begin with solar projects, so in this chapter, I introduce some ventures that get you out of the house. Operating a fountain with solar power, for example, means you can tune in to Mother Nature even more than with the usual solar projects. You can make fountains small enough to sit on a table or large enough to fill an entire yard. They sound soothing, and just about everybody likes waterfalls.

Or you can go low-tech, planting particular types of trees and shrubs in certain locations around your house to achieve dramatic improvements in temperature in both summer and winter. And with trees and bushes, you also get natural beauty.

Solar lights are easy and convenient functional decorations that anybody can install. You can make your yard come to life at night for less than $20. And for $200, you can light up the place like your own theme park.

Finally, controlling the amount of direct sunshine that falls on a patio or a porch enables you to spend a lot more time there. You can make it warmer in the winter and cooler in the summer. Think shade, the anti-solar solution!

## Building a Solar Fountain

Solar fountains are a simple, straightforward do-it-yourself project that work for any skill level. (The electrical work is bare bones minimum, with zero danger of shock or calamity.) Pumps and PV modules work well together.

You can literally connect the two wires from a PV module with the two wires of a DC pump motor, and you're in business. The pump works when you have sunshine, and the more, the merrier.

A basic fountain system has a minimum of several parts:

- A fountain (the structure, including the water path)
- A lower reservoir of water
- A pump
- Some tubing
- Wiring
- A power source

Some systems use an upper reservoir to store water, which evens out the erratic flow of the fountain on a partly cloudy day. In periods of bright sunshine, the upper reservoir fills up; when it's shady, it slowly drains down.

You can build a solar fountain any size you want, indoors or out, and you can create a soothing ambience with peaceful, gurgling water. Spend as little as $40 or as much as $10,000. All sizes are available for the do-it-yourselfer. The bigger jobs are just bigger, not more complex.

## *Making a basic barrel solar fountain*

An easy do-it-yourself fountain, shown in Figure 8-1, is made of two separate halves of oak barrels.

Here's how the fountain works: In sunlight, the upper reservoir fills slowly. When the sun goes behind a cloud, the fountain maintains flow by draining the upper reservoir. When the upper reservoir gets too full, water drains through the overflow spout. You can close the lower valve and save water in the upper reservoir for use later, such as when you go to bed or when you're having a dinner party with friends.

Water trickling sounds come from three sources: the valve, the overflow spout, and the pump tube. It's interesting to note the different sequences of sounds you get from various weather conditions and time of day.

A solar well pump system works on the same principles. During sunlight hours, well water is pumped into a reservoir located above a house. On sunny days, the reservoir fills up, but it has enough capacity to work even through a few cloudy days. Water runs into the house by simple gravity feed, with well-regulated pressure — at least until the reservoir drains.

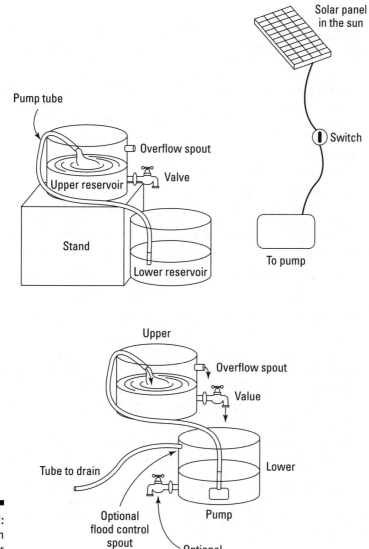

Pump tube

Solar panel in the sun

Overflow spout

Valve

Upper reservoir

Stand

Lower reservoir

Switch

To pump

Upper

Overflow spout

Value

Tube to drain

Lower

Optional flood control spout

Optional draw valve

Pump

**Figure 8-1:** A wooden barrel solar fountain.

## Parts list

You generally want to build the reservoir parts of the fountain before selecting a pump, so buy the materials in two stages. Here's what you need to build the basic structure of the fountain:

- ✔ **Two halves of an oak barrel:** These cost $20 and up per half at big nurseries. Nice barrels from wineries cost more, but they're very good quality and usually look better for a lot longer. They're not easy to cut in half, so buy them precut.

- ✔ **Clear silicon sealant:** Seal cracks and leaks with clear silicon sealant, sold at all hardware stores.

- ✔ **A completely reliable support structure:** Use heavy, rigid concrete blocks or bricks to support the upper reservoir, with the bottom around an inch higher than the top of the lower — it can get very heavy, so keep safety in mind. Keep the upper reservoir as level as possible. The lower barrel can sit on the floor, with the back edge supporting the upper reservoir.

The upper reservoir can weigh upward of 200 pounds if you're using big barrels. If you don't get it set with integrity, it can come crashing down. Always use concrete or mortar or glue whenever possible. If you need to, make the structure rigid enough so a kid can climb onto it without mishap.

- ✔ **A valve (faucet):** For $8 you can get a good brass faucet. Even better, gift shops often sell fancy faucets made of copper (green patina). Faucets made for outdoor hose systems come with male-threaded mounting ports, so you can screw the faucet into the barrel after you drill a hole.

- ✔ **One copper tube, around 4" long, between ½" and ¾":** Copper tubes are stock items in plumbing departments.

Or you can use a decorative valve that matches the other one as your overflow spout. Just open it up all the way and leave it like that.

After you build the basic fountain (see the next section) and before you buy the next set of materials, determine how much flow you need. Bring a hose over to the finished fountain, along with a 1-gallon container, and do the following:

1. **Place the hose in the 1-gallon container and adjust the water flow so that the hose fills the container in 1 minute.**

2. **Set the hose in the upper reservoir of your fountain and find out what kind of flow 1 gallon per minute results in.**

   If this is fine, buy a 1-gpm pump.

3. **If you're not satisfied with the result in 2, readjust the flow either up or down and measure the container fill time once again.**

   Use the formula: 1/fill time in minutes to get the flow rate.

4. **Reapply to the fountain and determine whether the new flow is correct.**

   The smaller the cheaper, but pumps can also wear out over time, so don't cut it too close. Many pumps have a flow adjustment valve; they're worth the extra money.

You're then ready to buy some of the more mechanical parts of the fountain:

- ✔ **A 12VDC submersible pump, at least ½ gpm (gallons per minute) flow with 2 feet of head pressure:** Head pressure is a pump spec, usually in feet, which tells you how much vertical height the pump will be able to force water. You may want bigger and faster, which makes more noise and moves more water. Most retail outlets for pumps assist their customers in choosing the best pump for the particular application. Use this free resource, if possible. Otherwise, the best bet is to look on the Internet.

- ✔ **Four feet of tubing that connects with the pump:** Clear plastic is the best choice because it's less visible. Buy matching tubing when you buy the pump (the pump output port will be a certain size; get tubing to match).

- ✔ **A 5-watt or larger PV module with 12VDC output:** How much power? Ask the pump experts, and they'll tell you how much flow you'll get for how much power. Keep in mind this flow is going to happen only under the best conditions.

  For $50 you can buy a complete kit with PV module, pump, wiring, and tubing. You don't get much flow, but it works just fine if you like subtlety.

- ✔ **A mounting scheme for the PV module:** You can lay a PV module directly on the ground or rocks or gravel, and it's not harmful. But in general, it does better the higher you can get it. Rooftop is best if you want a steady flow because you get sunshine all day without contending with shadows. If you want variation, plan to mount it in a place where shadows change over the course of the day.

- ✔ **Wire from the PV module to the pump:** Tell the hardware store clerk how big your solar module is in watts at 12VDC, and he should be able to give you the wire you need. Direct burial cable used for household 120VAC systems is a good option because you can simply dig a trench, lay it in, and then cover it back up with the dirt. It won't degrade over time.

  Your hardware store has a variety of wire splicing kits; when you're ready to do the wiring, make a diagram of where the pump and switch and PV module are; measure the distances in feet. Get both the wire you need and the proper splicing kit at the same time because they're one and the same.

✔ **An SPST (single pose, single throw) switch, electrical box, and face-plate:** The cheapest switch is an off/on light switch for a house. You can buy one for $1, and it's easy to wire. Buy a matching electrical box and faceplate; many styles are available. The most convenient boxes come with nails ready embedded into the plastic housing; hold it up to a post and pound the nails in, and you're hard mounted with integrity.

### Assembly and operation

Before you begin, get your first set of parts ready (see the preceding section for a list of what you need to build the fountain structure). Then follow these steps:

1. **Prepare the oak barrels and install the valve and overflow spout.**

   Seal any cracks and leaks in the barrels with clear silicon sealant. In one of the halves (the upper reservoir), drill a hole near the bottom for your valve (faucet) and install it. If your faucet has a male-threaded mounting port, you can twist it right in by hand (use clear silicon sealant very liberally and wipe away the excess after you have the valve in).

   In the same half, drill a hole for the overflow spout about an inch from the top, not directly over the valve but a couple of inches away. Angle the hole with a slightly downward tilt so when water drains through the spout, it'll pour outward.

2. **Finish the reservoir parts of the fountain.**

   Put the top and bottom reservoirs into place, setting the upper reservoir on its support structure with mortar, concrete, or glue.

3. **Determine how much flow you need and purchase the appropriate pump, tubing, and wiring (see the preceding "Parts list" section).**

4. **Install the solar pump system and put the tubing in place.**

   The pump goes at the bottom of the lower reservoir, as low as possible. The PV module goes into sunshine, and the tubing connects the pump to the upper reservoir.

5. **Complete the wiring, connecting the PV module to the switch and the switch to the pump.**

   Use your wire splicing kit. Make all wiring runs as short and direct as possible, and make excellent connections everywhere.

A shoddy connection is like a weak link in a chain; it defines the overall quality of the entire chain. Bad connections don't last long outdoors, in the elements. If you don't do a connection right, you'll be redoing it soon and suffering performance anxiety in the meantime.

Here's how to operate: With the switch off and the valve closed all the way, fill the upper reservoir one-quarter of the way. Fill the lower reservoir an inch from the top, and then flip the switch on. Adjust the valve so that the upper reservoir fills slowly on a sunny day.

Some tips:

- Oak barrels are made of individual wood staves, which are tightly pinched together by steel straps. If you let an oak barrel dry out, the gaps between the staves tend to warp and contract. Always keep the barrels as full as possible to avoid this issue. If it does become a problem, you can recover the original integrity by filling the barrels (which are going to leak) repeatedly until the leaks stop.

- When you buy your pond materials, get some anti-algae tablets. You can also use swimming pool chlorine, if it's handy.

You can have fun and discover a lot about solar by experimenting with all the different variables. A lot of subtle physics is at work — gravity, pumps, sunlight, weather, evaporation, balance, fluid flow, energy storage, power, and so on.

For example, you can face the solar panel east or west to change the time of day you want to run your fountain the strongest. If you like to wake up to a fountain, face the module due east. If you like to come home in the afternoon to a peaceful waterfall, face the PV module into the southwest. You can even use two PV modules connected in parallel to cover both early morning and late afternoon (when you buy the PV modules, make sure to tell the sales staff so they can get you the right kind to do this with). On the other hand, if you want your fountain to change a lot over the course of a day, put the solar module in a partly shaded backyard where shadows are constantly evolving.

## Designing your own master creation

All solar fountains have essentially the same types of operation and parts as the wooden barrel fountain. Designing and building your own version isn't hard. You can find a whole range of prefabricated fountains in gift shops and nurseries — check them for design ideas. If you have rocks, all you need is some mortar, and you can make one of the best fountain structures imaginable! Given a choice, I recommend the extra work of rock and mortar because you avoid leaks (with the oak barrels), and rock fountains tend to resist algae buildup better. Of course, a rock and mortar fountain is probably going to be fixed forever where you build it, whereas you can move an oak barrel fountain.

Your biggest decision is how large to make the upper reservoir. The larger the reservoir, the longer the capacity (in time) to continue flowing without sunlight on the solar panels. Then again, you don't need an upper reservoir at all — if you go without, your fountain goes exactly with the sunlight, and you can hear how much sunshine there is at any given time.

Deciding how much power you need impacts cost the most. For more power, you need both a bigger pump and a bigger PV module. Tall fountains require more expensive pumps and PV modules to move the same amount of water.

Try to go for subtle. If you want to make up for lack of flow, make a lot of little fountains instead of one big one. Make your fountain shallow and broad. The water levels in narrow, deep fountains change quite a bit, which can be unattractive and cause big fluctuations in the flow levels.

Many landscape fountains use a black rubber liner that fits snugly over almost any contour. It's easy to install, seals very well, and lasts a long time. Unleash your imagination. Broad, shallow reservoirs work better than narrow, deep ones because the water levels don't vary nearly as much.

You have to keep your solar fountain topped off with water because you can't let the pump run dry. The easiest way is to connect a drip valve from a landscape watering system into the lower reservoir (just barely suspend the dripper over the edge of the lower reservoir — don't let it hang down into the water, or it may siphon out your water). Make sure to leave a drainage path for when the lower reservoir overflows (you don't want to use this method if your fountain is indoors or sitting on a finished floor such as tile). If you're not getting enough water to keep the lower reservoir filled, use more dripper heads. Overflow doesn't hurt the system, but you don't want the pump to run dry, which happens when you don't have enough water. Err on the side of full.

The best fountains blend in with the natural scenery. Use natural rock and surround the fountain with plants. If you're interested in creating a natural ecosystem and using a solar pump to enliven it with some moving water, use broad, shallow reservoirs, and locate the valve only an inch or so below the level of the overflow spout. The depths of the ponds will vary only by that amount.

You can build a good-sized solar fountain for $300 if you use natural materials and your own elbow grease. Just $500 can get you a beautiful koi pond fed by a naturally flowing spring — a solar spring, that is!

If you want to do an installation but not a design, buy a complete fountain or pond kit, which comes with detailed instructions and expert suggestions. Each kit comes with specific design suggestions, including a parts list and labor instructions. Or you can buy a prefabricated fountain without a pump and then install your own solar components. Look in the Yellow Pages under fountains and ponds.

# Lighting Your Yard with Solar

One of the easiest and cheapest ways to start out in solar is by installing lighting in your landscape. For $15, you can get a decent solar light with a range of mounting schemes. The simplest units come with a built-in stake so you literally don't do anything more than stick it into the ground. If it turns out that you don't like that spot, pull it out and stick it somewhere else.

The vast majority of solar landscaping lights come one-piece, with the PV module on top of the light itself, so you should put the whole thing where it can get a reasonable amount of direct sunlight. However, solar lights may surprise you by how well they work, given a meager amount of light. If the lights don't get much sun, they still come on at dusk — they just don't last all the way to dawn. Your best bet is to try one out in the location you want, even if no direct sunlight is there. It may work just fine.

Some lights are static, meaning they don't blink or change colors. Background lighting, for example, should be static; it should establish a sense of place and highlight the best features of the environment. The most functional locations for the spotlight variety are around porches and walkways and along driveways where people will be walking. Put one near your garbage can out back, and you don't have to flip the light switch anymore. You can also get solar lights connected to motion detectors (see Chapter 9).

Other lights revolve through patterns of color and brightness. The effect is entirely different. Dynamic lights should add a subtle hint of presence. One of my favorite solar lights is a clear plastic butterfly that changes colors slowly and subtly. It comes on a stake with a 2-foot wand to the butterfly, which seems to be floating because you can't see the wand at night. It draws your eye without demanding it, and the colors are rich and textured.

Don't buy cheap lights, because they don't last. If it's in a flimsy plastic housing, don't buy it. Aluminum is good; heavy black plastic is cheaper and works just as well. You may want to buy an entire matching set; 6 units for $60 is usually good quality.

# Planting Trees and Shrubs for Shade and Wind Control

Plants add natural beauty and keep the air supply clean by trapping dust, consuming carbon, and making oxygen. But adding some leafy vegetation can give you some other perks as well.

With growth fueled by the sun, plant cells are Mother Nature's original solar cells — and you can harness the power of landscaping to provide shade and keep you cool. By planting deciduous trees and shrubs, you can work the sun to your advantage in both summer and winter. The leaves provide shade in the summer, and the bare branches let sunlight in during cold winter months. And by planting trees around your house, you can also increase or decrease the breezes through the house. This section tells you how.

## Planting for your day in the shade

*Thermal mass* is the amount of heat energy a substance can hold. Bricks, concrete, stones, and the like hold a lot of heat. Wooden decks are pretty good at it as well. So if the sun shines on a concrete patio, that patio is going to be hot most of the night. But landscape bark, mulch, and plants don't hold much heat at all. If sun shines on landscape bark all day, there won't be much stored energy at the end of the day, and the temperature goes down reasonably fast. So a bit of strategic planting can help you create some cool spots.

A wall or roof that's shaded from a tree or shrub can be 20°F cooler than the same thing in full sun. On a hot summer day, the air temperature under a tree can be 25°F cooler than on nearby blacktop.

Big windows to the south of your house are the best location for shade planting. Sliding glass doors or big, wide doors of any kind are best of all. Also plant for summer shade over high-thermal-mass patios, porches, sidewalks, wooden and synthetic decks, and so on. The closer to the house, the better. Choosing deciduous plants — which shed their leaves in the fall — ensures summer shade and winter warming.

## Directing the wind

Figure 8-2 shows you how to plant trees to either decrease or increase the breezes through your house.

Your *prevailing wind* is the direction the wind blows most of the time. (Go outside, observe the direction of the wind, and then keep a log over a period of time — say, ten measurements a month over the course of a year.) To decide where to plant, draw an overhead view of your house and then include an arrow for prevailing wind. Mark the location of each window, and then plan accordingly. The prevailing wind may change over the course of the seasons, so you'll have to decide which season is the most important in your operating scheme. In general, a good design includes plants for both winter and summer.

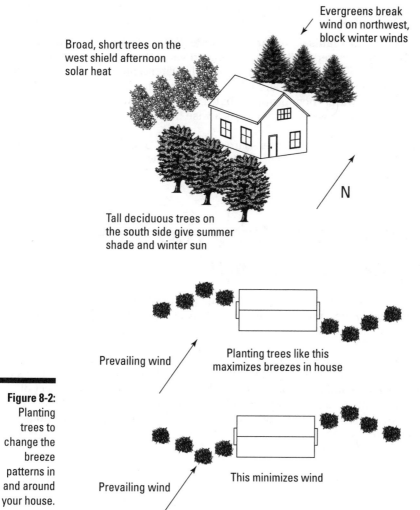

Evergreens break wind on northwest, block winter winds

Broad, short trees on the west shield afternoon solar heat

Tall deciduous trees on the south side give summer shade and winter sun

N

Prevailing wind

Planting trees like this maximizes breezes in house

Prevailing wind

This minimizes wind

**Figure 8-2:** Planting trees to change the breeze patterns in and around your house.

Here are some tips on plant selection and care when planting windbreaks:

- **If you want the most tree the fastest, don't get the biggest tree — get the one with the biggest root ball.** Make sure that the nursery didn't just plant a small root ball into a big container. The bigger the root ball (if it's healthy), the bigger your tree will be in a few years.

- **Regardless of where you live, use natural and indigenous plants.** They grow better and are proven to last. Talk to the staff at a local nursery about your project and show them your drawing. They should know the perfect plants for each spot, and they can tell you how to get the fastest growth with fertilizers and water scheduling. In general, you want evergreens for winter windbreaks, but your options for summer are wide open.

- **Keep your plants on a water schedule.** For about $20, you can get a battery-powered digitally programmed valve that works great anywhere you have a hose. You can automatically water a number of trees for $40, including the drippers and tubes. Trees and shrubs grow like mad if you feed and water them properly, but they don't grow much at all if you don't. Making water automatic is a wise investment.

Convection cooling (see Chapter 4) from the wind is a big heat loss in swimming pools. By placing a perimeter of shrubs on the windward side of the pool, you can make your pool look better and swim in warmer water at the same time.

# Anti-solar Power: Using Awnings, Lattices, and Overhangs

Sometimes you don't want to harness the power of the sun so much as control it. This section tells you a bit about ways to tap into the power of shade so you can keep your outdoor areas comfortably cool.

The most beneficial spot to shade your house with outdoor blinds and screens is a back porch, especially if it's concrete or stone. Wooden decks are also good candidates because they hold a lot of heat. The net effect of making a patio or a porch more comfortable is that you can spend more and better time there. You basically increase the square footage of your house with this investment!

You can also build a nice looking lattice structure overhead that acts as a sun break. Cover the lattice with a sunscreen to keep heat and bugs out at the same time. Or even better, use deciduous vines. Be careful with them, however; some (like grapes) are savagely invasive and will take over your entire yard in a year or two. Of course, this may be your goal. (Grapes also yield fruit, which is very nice.)

If you screen in a covered porch, you can leave the doors and windows into the house wide open. Not only can this make you feel like you have an extra room in the house, but the house will be considerably cooler as well. You can either build your own roof or buy kits for roofs made of canvas or vinyl or fiberglass (see Chapter 15.)

Automatic awnings (motor-powered, either solar, battery, or household current) are also available — they work via a handheld remote control (the same kind used on TVs, so they're very easy to control).

Gazebos are available in complete kits to cover anywhere from 10 square feet to over a few hundred. In hot climates, a gazebo often makes the difference between rarely using a patio or lawn space and always using it. High-quality units come with screens that you can easily open and close. Some units come with solar ceiling fans. You can spend anywhere from $50 to tens of thousands of dollars on complete units.

Overhangs are often good do-it-yourself projects with both financial gain and aesthetic beauty. Figure 8-3 shows the basic idea.

You can design an overhang to match your house and roofline. This can make a huge difference in the appearance of your house and give you function at the same time.

You'll have to design your own overhang because each one is so different that no universal designs or supplies exist. The best bet is to use the same materials your house and roof are made of. Make every effort to make the overhang look like it belongs to your house. Before you start nailing things down, set up the materials in the best approximation of the final design and see what it looks like.

This is not a simple do-it-yourself project because the design phase is very important. On the other hand, if your design is simple, most people can build an overhang with minimal tools and experience.

You can get a carpenter to help you, and it shouldn't cost too much. You can get a carpenter to do only the design for you, and you can build the design. Go to a construction site and just ask one of the framers. You don't need a licensed contractor because you don't need a license.

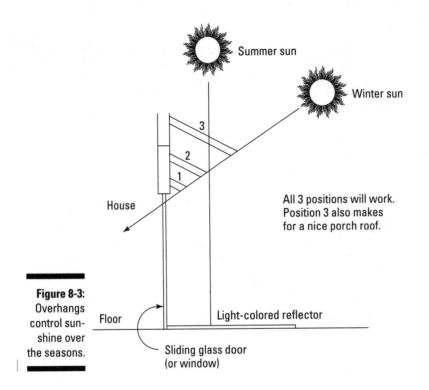

Summer sun

Winter sun

3

2

1

House

All 3 positions will work.
Position 3 also makes
for a nice porch roof.

**Figure 8-3:**
Overhangs
control sun-
shine over
the seasons.

Floor

Light-colored reflector

Sliding glass door
(or window)

# Chapter 9

# Jumping In to Small Projects with Big Results

*S*tarting out with small projects makes sense because you find out how solar works through experience. When you move up to bigger and more powerful systems, you'll better understand the ropes. In payback terminology, you decrease risk by starting out simply.

In this chapter, I present all kinds of fast and easy solar projects. Some are sensible; some are just for fun. Who's to say which is more important?

## Going for Gizmos, Gadgets, and More

Whatever you choose for your first solar experiments, make sure that you have fun with them. You may want to begin with some of these solar toys and novelties:

✔ **Solar clip fans for hats:** Cooler heads will prevail! Hat fans clip on to your hat visor, and a tiny PV cell spins a fan, which you can aim at your face. At around $10 apiece, they're great for baseball games, the beach, reading, or napping. The lifetime of the product is speculative at best, but you can have fun in the meantime. Buy a batch and give one to each kid you know, or give them out at promotions — everybody will remember you.

- **Spinning prism:** This $20 toy attaches to your window via a rubber suction cup. Facing outdoors is a PV panel that feeds a motor that spins a large crystal. When the sun shines directly onto the PV panel, the crystal sends twirling prismic patterns around the room. At some point, you may get weary of it — kids don't, though.

- **Solar dollhouse:** PV panels on the roof can power hair drying and other necessities of modern life — all for zero pollution. Be prepared when the young owner wants to know why the lights work only when you don't really need them.

- **Solar car:** With PV cells on this car's roof, the sunnier the faster. That's a curious performance characteristic for a car — not really practical, but hey.

- **Solar pond fountain:** The solar pond fountain is a one-piece float for your swimming pool or pond. Solar panels power a pump that squirts a little fountain straight up — the more sunshine, the more boisterous. Just take it out of the box and float it. Some look like lily pads; very cheesy. You can typically pick one up for $30.

If you want your gadgets a little more practical (at least in theory), you may want to consider the items I describe in the following sections.

## Washing with portable showers

With the solar portable shower, you fill a specially constructed plastic bag with water and then place it in direct sunlight to heat up. The 5-gallon bag is clear on top, facing the sun while it's warming. The back wall is black to absorb maximum sunlight. Most portable showers come with a thermometer so you know when it's ready.

At that point, hang it from a tree for a gravity-fed, hot shower. It's convenient for camping and backpacking. It's also good on boats when you want to get cleaned off at the end of the day. You can hang one out by your swimming pool for a quick, simple rinse. They can also serve as solar hot water bottles — big, hot, soft, and free.

If you forget about them, portable showers can get really hot, over 120°F, so beware. Always sample the temperature before you step in.

## Shining light on solar flashlights

A solar flashlight always has fresh batteries — as long as you don't store it in a dark drawer or cabinet. Get one of these and store it on the windowsill.

They're bright enough to be functional, and they last a long time after they're charged. Are they worth $25 (compared to conventional flashlights for $10)? Well, they're cool, and because you have to set them out in the open, they're conversation pieces. And if you're the type who always runs out of batteries, this may be the cure. Seriously, these are very practical and they last forever; they're well worth the $20 cost.

## Doing yard work with solar machinery

You can buy a solar/battery-powered lawnmower — but you can't store it in a dark shed, or it won't do much work! With a full charge, you can mow for around an hour, which should cover even the biggest lawn. They weigh about as much as a regular mower, but do cost more. They work the same way; you have to sweat and grunt and groan and push them over every square inch, so you may just want to pay a neighbor kid to the do the job for you.

You can buy a solar-powered leaf blower, which makes sense for small yards where gas is overkill, electric power cords are a hassle, and conventional rechargeable units have to be located near an AC plug. When not in use, you can store a solar-powered leaf blower anywhere you have sunlight.

## Creating ambience with swimming pool novelty lights

Here's my favorite toy: a clear plastic disk, around 6 inches in diameter and 2 inches thick that floats in a pool. During the day, a battery charges via a PV module. At night, an eight-color LED display cycles through an evolving pattern that shines down into the water so that the entire pool glows with the color. These lights can sauce up your pool like nothing else. Static lights create ambience, but moving lights create presence.

They're cheap — and they should be because they don't last long. But at $12 apiece, it's okay. If you have a pool, you know $12 is nothing in the grand scheme of things.

Use at least two lights at a time — they have the same patterns but different speeds. Sometimes they get in sync. Four work best.

I recommend getting some for a party. Make sure that they're in plenty of sunlight before the party starts. Tell kids they're spaceships, and they'll probably believe you!

## Spreading holiday cheer

Running long, snarled power cords to your strings of Christmas lights is a thing of the past. Now you can use solar-powered lights that don't plug in at all. A small PV module charges a battery during the day. At night, illumination comes from low-power LEDs. But they show up loud and clear when they're aimed right. If you want to put up only a string or two, the job takes about 10 minutes. Solar works very well in freezing, clear weather, and the LED lights look great against snow.

You may not get a lot of Christmas cheer in extended periods of thick weather. The lights do always go on after dusk for at least some time, but if they didn't get much charge, they go off soon. Still, do you really need lights after midnight?

# Coverings and Barriers: Letting (Or Not Letting) the Sun Shine In

Controlling the amount of sunlight that enters your house or patio is an easy way to both regulate temperature and enhance the appearance of your house.

## Choosing locations for window coverings

Window coverings are a great way to control the amount of light that comes in and to create an insulating or cooling effect in your home. Figure 9-1 shows your different window covering options, and Table 9-1 explains what happens with each one.

With a blind inside, you can get the best insulation. However, the heat gets inside because sunshine comes in through the window and strikes the blind. The gap between the blind and the window can get very hot because of the greenhouse effect. This result is desirable in the winter but not the summer, when you may have a cooler room if you just leave the window uncovered and let the sun and air in.

You can get interior blinds made to warm a room in the winter if that's your main goal. They absorb sunlight, and they also provide insulation where you need it. On the other hand, interior blinds for summer reflect a lot of light on the outside (exterior) surface, which may be made of a different material or made a different color, from the inside surface. In general, a white blind works well for reflecting sunlight the best, but you can also find (more expensive) specially coated blinds that reflect the maximum amount of sunlight.

With a solar screen tacked up on the outside of a window, most of the sunlight is filtered before it even gets to the window. The screen gets hot, and you end up with an insulation barrier to keep that heat trapped, but it's all outside the window, which makes for a cooler interior. This is good for summertime, but counterproductive for winter.

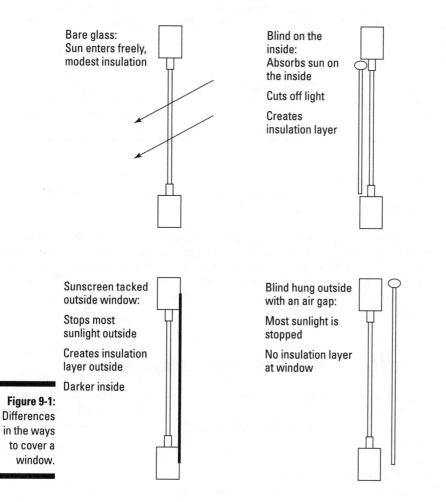

Bare glass:
Sun enters freely,
modest insulation

Blind on the
inside:
Absorbs sun on
the inside

Cuts off light

Creates
insulation layer

Sunscreen tacked
outside window:

Stops most
sunlight outside

Creates insulation
layer outside

Darker inside

Blind hung outside
with an air gap:

Most sunlight is
stopped

No insulation layer
at window

**Figure 9-1:**
Differences
in the ways
to cover a
window.

| Table 9-1 | Window Covering Options | |
|---|---|---|
| *Type of Covering* | *Sun* | *Insulation* |
| Bare glass | Sun enters freely | Minimal insulation |
| Blind on the inside | Absorbs sun on the inside; controls lighting and enhances decorative qualities in the room; may also completely block view | Creates excellent insulation layer inside — good for cold climates or winter |
| Sunscreen tacked outside window | Stops most sunlight outside; darker inside but with partially obscured views | Creates insulation layer outside; heat is stopped outside so that the interior of the room is cooler; best for hot climates |
| Blind hung outside with an air gap | Stops most sunlight outside, partially obscured views | No insulation layer at window; maximum cooling effect, especially in breezy locations |

The best cooling effect is to allow for some ventilation between an outside blind and the opened window that it's covering. But that setup isn't always easy or practical.

When your air conditioner is on, you want to create as much insulation as possible. You can do so by closing your blinds all the way. If you have outside covers, lower them into place. The best scheme for a sunny house that uses a lot of air conditioning is both solar screens and interior blinds. The house is relatively dark inside, but it's also cooler. (You can also use window overhangs — see Chapter 8.)

## Investing in blinds and drapes

Windows are the showpiece in any room, and window covers make a big difference in the decor. With a stock blind from a hardware supplier, you get mounting hardware and instructions, plus quality that'll last for more than a few seasons.

### Indoor blinds and drapes

You can get the most bang for your decorating buck with inside blinds, spending anywhere from $10 to thousands. Inside blinds and drapes are good for cold weather. The best insulators are made of honeycombed fabrics. Keep them open whenever you have sun; close them otherwise, especially on a cold night. Curtains work the same way, and you have a much broader selection of fabrics and styles. In the summer, you may not even notice the difference between open and closed blinds and drapes.

### Outdoor blinds

You can easily hang an outdoor blind on the frame above a window or a door. Some outdoor blinds come with sunscreen (see the preceding section).

Outdoor blinds are usually made of vinyl. When you hang one outside your window, that's what you're going to be seeing from the inside. Some are prettier than others.

If you want to raise and lower the blind often, retraction should be convenient. Most blinds are manual, with cords that you pull different ways to change the setup. Some are automatic, and you can even find solar-powered retractors that work with a hand-held remote control. You don't even need to go outside.

Import shops stock different sizes of rollup bamboo shades, often as low as $5. These shades work well outside because they look reasonably good and are cheap enough so that when the weather eats it up, you can just toss it out. You need to get some mounting hardware, probably in a different store (unless you just nail the shades up there with whatever's on hand, which can be minimum cost for maximum gain).

## Installing sunscreens for summer

Installing sunscreens is the cheapest way to prepare a window for the summer season, and it's also one of the best in terms of practical effect. Sunscreens reflect a lot of sunlight and create an insulation barrier on the outside.

Most sunscreens are dark, heavy-duty fabric screens with a shiny surface. Some are a flexible, tinted plastic film. They can reflect up to 90 percent of all sunlight, which makes a big difference inside the house.

### Looking at placement, payback, and installation advice

Only windows that get more than a few hours of direct sunlight a day benefit from sunscreen installation. The best candidates are tall windows facing south, but east and west can also get very hot. Of course, part of the effect is the psychological comfort of shading, and you may want that anywhere.

You don't want sunscreens in the winter, so they're only temporary. The solution is to mount them in removable frames or tack them up. You can also roll them up like a blind or buy a finished blind that has solar screen. Or get automatic, electric controllers so that with the push of a single button, you can command every solar screen in your house.

You can expect at least three years' lifetime with quality sunscreen, at around $0.75 per square foot. (Cheap screen bleaches out and looks perpetually dirty, so spend a little more.) For a 3' x 6' window, if a screen lasts five years, the cost is only $2 per year (with the cheapest installation method, namely tacking it up with thumbtacks). If your climate is hot and sunny, you'll undoubtedly get good payback on this small investment.

Call a screen shop directly for tools, materials, and advice on how to install sunscreen. Sunscreens come up to 8 feet wide, in huge rolls, so that you can get as much length as you want. You can do the installation with nothing more than tacks, a hammer, and a box cutter, but a good stapler ensures a quality, consistent job and is much easier.

You may want to ask the shop about coming out and putting up the sunscreen for you. The workers can do a window in a few minutes, and their bid for the entire job may not be much different from what you'd pay for the material alone.

### Getting framed

You can buy frames for sunscreens in a range of colors, and they come either loaded with screen or empty. Most are aluminum. Easy-to-use mounting hardware is an essential. Buy that when you get the frames because it's one and the same problem; the frames are useless without mounting hardware.

## Creating a manual screen retractor

The figure shows a way to make a manually operated screen retractor that works reliably, is easy to use, and costs very little. You need a wooden mount that's slightly wider than your window, three eyelets, a sunscreen, tacks, a dowel rod (as long as the mount), and screws or nails (to install).

1. **Drill three holes in the mount and attach the eyelets.**

   The first two holes are 6" from each end of the mount; the remaining hole should be 1" from the end. Screw in the eyelets, making sure that all three holes are lined up, facing the ends of the mount.

2. **Using a box cutter, cut the sunscreen to size.**

   You want to make the sunscreen larger than the window by about 4 inches, although

if you don't have the room, any size will work, even if it's smaller than the window. A size smaller than the window will just be unsightly from inside, looking outside.

3. **Tack the sunscreen onto the roll-up rod (dowel) and the mount.**

4. **Put the cords in place.**

   Tie a cord to an eyelet that's 6" from the end, loop the cord over the top of the mount and under the rod, and go back up through the eyelet. Repeat on the other side. Then thread both cords through the end eyelet, the one that's 1" from the end of the mount.

5. **Screw or nail the mount onto the window frame.**

   To retract the screen, simply pull the cords.

If you like, paint the wood to match your house. You can use pulleys instead of eyelets and get smoother performance. If wind is an issue, get some fishing weights (the kind with holes so you can get a nail through) and nail them into the ends of the roller dowel, equal weight on each side.

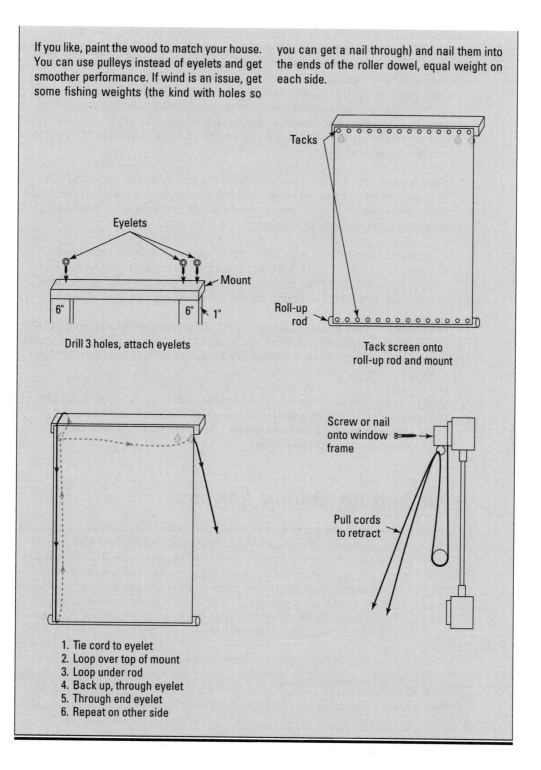

Eyelets

Mount

6"    6"    1"

Drill 3 holes, attach eyelets

Tacks

Roll-up rod

Tack screen onto roll-up rod and mount

1. Tie cord to eyelet
2. Loop over top of mount
3. Loop under rod
4. Back up, through eyelet
5. Through end eyelet
6. Repeat on other side

Screw or nail onto window frame

Pull cords to retract

Houses with wooden window frames work the best because they accept nearly every mounting scheme. If your house is stucco or you don't have wood around your windows, you're limited to prefabricated frames with appropriate mounting hardware. The job is more expensive but still potentially worthwhile. Calculating the cost savings is difficult, so you'll have to use your budgetary constraints to decide whether it's worthwhile. To find out what the effect will be, get some cheap screen and tape it up to see how it changes your room; you may not like the effect, which will govern how much you're willing to spend.

If the frame's wood, the fastest and cheapest way to cover a window is to tack the screen right onto the window frame, using a hammer and tacks. Cut the piece of sunscreen down to a few inches bigger than what you need, tack it up, and then trim it with a scissors or a box cutter.

If you want a better look, consider cutting some planks of wood to frame the window, painting them, tacking them up over the screen (the screen is already tacked up over the window — the frames are only for appearance, not functionality), and trimming the excess screen to match.

Get the wood from a lumber store — you can probably find a store that cuts the wood for you, and the pieces will be exact and square. (Cabinet shops are usually willing to do a small cutting job if you can't find a lumber store that does it.)

Using screws for installation is best because taking the screens down and then putting them back up next year is easier. You need a drill to predrill holes for the screws, and a screwdriver. The lumber store staff can show you the best screws and drill bits to use.

## Putting up radiant barriers

A *radiant barrier* is a sheet of thin material that looks like reinforced aluminum foil. You usually tack it up beneath your roof joists, and it keeps a lot of heat out of your attic. The material itself isn't all that expensive at $0.20 per square foot. For a 2,000-square-foot house, the material costs $400. (Because your roof joists are angled, you need more material than the square footage of your house. Multiply your square footage by 1.25 to get a good approximation of how much you need. Don't worry if you don't cover the entire roof. Also, do not cover up any vents because radiant barriers will not transmit air.)

These barriers are advisable in hot climates because they cool down the house more than enough to pay you back in a few years. But if you want your attic to get warm in the winter, they may not be the best solution because they prevent heat from entering, resulting in a cooler house. Controlling the temperature in your attic is always a balance between seasonal extremes, so decide whether heat or cold is your major problem.

Installation is a bear, unless you can lay the material directly onto the floor (or over the joists and insulation comprising the floor). Open rafters are the best candidates, but if you have no floor and you have to hop around on joists, you may want to find a different solar investment.

When installing on overhead joists, absolutely get a very good stapler. Electric is best — get a powerful one. You're going to be holding your arms over your head half the time, and then you'll be crouching over and twisting into a pretzel shape the other half. When you pull the trigger, you want a good, solid staple to go in without argument.

Even if you can cover only a portion of the rafters, radiant barriers are worth it. Try to cover an entire medium sized area rather than small spots at a number of different areas. The hottest spots on your roof (usually the southern exposures) are the best candidates.

Walls also work with radiant barriers, and doing a wall is usually a lot easier. An uninsulated garage that gets way too hot in the summer is a good candidate, although it may look like a spaceship when you're done.

Before you decide to spend the money and install a radiant barrier, go up into your attic and see what it's going to be like crawling around nailing it up. Envision lifting the material up into the attic and then unrolling it, cutting it, and holding it up into place. Envision the tools you need to carry around. One person can't put up radiant barriers, so get some assistance.

Attics can roast you fast. They can heat to over 130°F, easy. Working in the morning usually solves this problem, as does working in cooler months. Don't kid yourself; if the temperature feels really hot, it can be dangerous.

# Securing Your Home

You can make some easy, solar-powered security systems for very little cost. You can also have a lot of fun at somebody else's expense. By using strategic combinations of motion-activated sprinkler heads and security lights, you can keep just about anything or anybody away from your house. The great news is that both of these products come solar-powered, so they work even in a power outage. These security devices are far from foolproof, but they deter 99 percent of potential offenders. The trick is placement. Plan for maximum surprise and dramatic result.

## Hosing down your problems

The idea is simple: Aim a stream of water into a region being sensed by a motion detector — upon detection, the water triggers on. The entire thing is powered by a solar panel, so you can set it up anywhere you have a hose.

Here are some intriguing applications:

- **Security:** If I were a burglar stalking a house and all of a sudden I was hosed, I'd leave pronto.

- **Pest deterrence:** Neighbor's dog like your yard late at night? Deer eating your landscape? Daughter's boyfriend pushing it? Not if they have to take an unplanned shower.

- **Practical jokes:** You can get some laughs out of this, but you better make sure that the intended victim has a sense of humor and is dressed appropriately. Set up your video camera first.

## Blasting the bad guys with light

The device is called a security light, but you can find other applications. A PV module connects via a length of wire to a battery, a spotlight, and a motion detector. The module needs to be mounted in direct sunlight, the more the better, but you can mount the light itself anywhere. Wire lengths of 15 feet are common, but you can go a lot longer if you need to. The typical price is around $60.

On the other hand, if you pay an electrician to install an electric junction box, you're looking at over $400, easy. And that location may not really be the best location after all. With solar, you can position anywhere, easily and quickly. And you can hang one of these inside your house just as well as out. The trick is to get the PV panel in the sun — that's all that matters.

Here are some applications:

- Aim it out over your front porch, and nobody ever has to approach your house in the dark. Or mount it over your garage door, facing the driveway, to provide light whenever somebody drives up.

- Aim it at a door, and whoever comes through suddenly finds himself or herself in a spotlight.

- Use it as a motion-activated light inside your garage, or hang the light over the work area in a remote storage shed.

- Put up a barn light to welcome the cows when they finally come home.

✔ Dark basements are perfect. When you go down, the light comes on. It goes off after you leave — no more grabbing around for switches and burned-out light bulbs.

✔ Get some gazebo lighting or position one over a picnic table. Whenever you sit down, you have light.

✔ In a remote cabin, you can provide all your lighting needs, indoors and out. Just remember, motion sensors have a preset on time (you can adjust this with some units to over ten minutes) so mount the units where they'll consistently pick up motion in the room. If not, you may be caught in the dark, waving your hands around in an attempt to shed some light. Actually, that sounds kind of fun — you may even be able to invent a game out of it.

# Engaging in Small Power Projects

You can run just about anything off of PV panels. The biggest questions are

✔ **Whether or not you need a battery:** A battery is required whenever you'll be using a device when the sun is not shining, such as at night.

✔ **Whether you need an inverter:** PV panels put out DC voltages only; if you need AC voltages, such as the kind that appliances take, you need an inverter.

## Battery power: Charging up your life

PV modules are ideal for charging batteries. The electronics are minimal, and the costs are low because of it.

### Leaving dead car batteries behind

Got an old truck, car, or RV that sits around and doesn't get started very often? How about a boat? If they've been giving you problems starting, a solar charger may be a very cheap and effective solution. When a vehicle sits around without being started up for a while, its battery grows weak. If the battery is old, it's even worse.

For $40, you can get a solar battery charger. A PV module plugs in to the cigarette lighter jack, and you lay the module out on the dashboard in the sun. While you're gone, it trickle charges your battery. A solar charger won't overcharge your battery, so you don't have to worry about removing it, even when the vehicle is running. But it probably won't revive a dead battery, although it won't hurt anything to try.

Auto batteries cost around $80. With the $40 solar charger, you get more lifetime, up to 50 percent more, so there's a reasonable payback. But how much is it worth (in peace of mind, not to mention the costs of a tow-truck visit) not to get stranded when you can't get your car started?

### Charging your electronics batteries with the sun

Standard, off-the-shelf batteries cost around $0.75 apiece. The cost for 100 throw-away alkaline batteries is $75, plus trips to the store to buy them. And batteries have nasty chemicals, such as lead and acids. You run across an environmental issue with 100 batteries.

Rechargeable batteries cost $4 apiece, and the good ones issue the same charge as a throw-away. A solar charger costs $40, but the charge cycles are free. You can charge a good rechargeable battery over 500 times. (Note that the number 500 applies to devices like remote controllers, which don't take much current; if you're drawing a lot of current and running the batteries down to their minimum, expect more on the order of 200 times.)

The cost for a rechargeable battery and 100 charges is $44, which is already a better deal than alkaline. But here's the best part: The cost to charge the next 100 times is zero. And the next, and the next. After 500 free charges, you may need a new battery, but that's only $4.

Alkalines cost about 20 times as much as quality rechargeables to operate. (Cheap rechargeables have much worse payback, so avoid them.) Spend some extra money on quality batteries, and it's a good investment.

The time to charge batteries (most devices charge four at a time) depends on how much direct sunlight you receive. Setting up near a window is often good enough if you don't need a lot of batteries. If you use a lot, you need direct sunlight. Kitchen bay windows are convenient candidates.

Keep a reserve of charged batteries. Buy twice as many as you need — it doesn't cost more because you go twice as long before you need to buy new batteries.

## Small-scale PV systems: Using just a modest amount of power

For $1,000, you can get a solar power system big enough to run anything in your house aside from the major appliances. You can power a home office, where you may be able to write off some costs as a business expense. Or use a solar power system in a remote cabin or a boat cabin or an RV. (For info on large-scale PV systems, see Part IV.)

About $1,000 can get you 500 watt hours, or ½ kWh, or a 60-watt light bulb for 8 hours. This is plenty of power to run an office or all the equipment other than power tools at a construction site. At this size (on the order of a small end table), you won't be toting your solar system around in a case, although you can find units on sturdy wheels.

## Working with RV appliances

Small appliances for RVs, boats, and camping are widely available, and most of these work with 12VDC solar power systems.

You may even be able to run some appliances without batteries, which makes things a lot cheaper and easier. Heaters don't care how the power looks (it can fluctuate, the voltages and currents can change, there can be ripples galore); how much heat you can generate is all that matters. Fans may blow directly from a PV module; the more sun, the harder the fan pushes. Water pumps don't need batteries, and some portable coolers are made to connect directly to solar panels.

You can also get a power system with a 120VAC outlet so you can use your existing household electrical devices. You give up efficiency, but it may be worth it. If you have a 12VDC system, you can buy a really cheap 120VAC inverter from an auto supply store. They come with a cigarette lighter plug, so you may have to do some splicing because most solar systems don't come with cigarette lighter jacks. The AC power is ragged, but it works for certain things just fine. It's difficult to specify a list for what will work and what won't without giving it a try. Radios may be noisy, as well as TVs, but some have good internal power regulators. Anything with an internal battery won't mind the noisy AC power.

# Going Off-Grid on a Piece-by-Piece Basis

You can go off-grid with small functions — you don't need to go off-grid with your entire home. You'll save on your power bill, plus enjoy some interesting independence and help save the environment, all in one.

## Making your reading lamp go off-grid

If you like to read, you can spend $50 to get a small, battery-charged light (LED) that works for four hours on a four-hour charge. A PV module with a length of wire attaches to a battery/light/switch housing with Velcro backing.

Apply the matching Velcro anywhere you may want light. During the day, plug in the PV module and set that in the direct sunlight. At night, press the light into the Velcro and use it.

Solar lamps put out a lot of light, but it's focused on a spot. The trick is to get that spot on your book, and you'll find that you have plenty of light. For this, you want a lamp with an arm that extends out, over your book.

Stick Velcro onto the bottom of a conventional lamp head employing a gooseneck adjustment arm, and press the solar light into that. You'll be able to position it directly over your book, about a foot and a half away.

If you're using the solar lamp only in one location, run a cord over to a window and tape the solar module flat onto the inside, facing out. You'll probably get enough sunlight to meet your needs. If not, look for more-direct sunlight.

If it's not bright enough, use two lights. You can also buy more-expensive designs with better lights, color, focusing, and battery.

## Installing solar light tubes (tubular skylights)

A *solar tube* lighting system collects sunlight on the roof and transmits it down a shiny, silver pipe into the diffuser, which broadcasts the light into the room below. Figure 9-2 shows how a solar tube lighting system works. You can make most rooms bright enough to work in, and they stay a lot cooler than if you use conventional lighting, like incandescents.

Solar light varies with the clouds and weather, changing the intensity of light in the room quite a bit. You're much more conscious of the outdoors. When the sky is partly cloudy, you can get a lot of fluctuation. Personally, this is my favorite aspect of solar tubes but some people find it objectionable.

The light creates a certain cool mood due to the silver color. Some light tubes come with filters for creating moods, but they cut out light as well. You can change the nature of a room very dramatically with a filter.

### Costs and payback

The typical price (uninstalled) is $210 for a 4-foot pipe length. After that, you need to buy extensions for $20 a foot.

Large-diameter units can output as much light as a dozen 100-watt light bulbs at one-tenth the heat. If a $250 solar tube displaces only three 100-watt light bulbs for eight hours a day, it pays for itself in 19 months. If you factor in the cooling effect, payback is even faster.

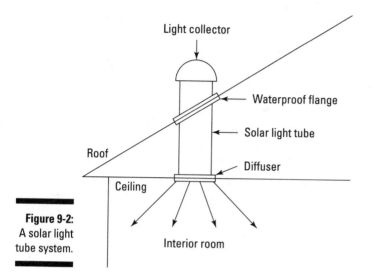

**Figure 9-2:**
A solar light tube system.

Light collector

Waterproof flange

Solar light tube

Roof

Diffuser

Ceiling

Interior room

## Location

The best locations are dark corners in family rooms. That way, the light gets used the most and has the most dramatic affect. They're also good for dark, isolated bathrooms. The natural light is comforting, and you never have to flip a light switch during the day. Beware, though. They can drastically change the way decor looks, maybe making the existing facades obsolete. Be prepared for a much different bathroom.

If you have a dark kitchen, a solar tube may be the perfect solution, especially if a lot of people come and go during the day. The cool tone of the light goes especially well in a stainless steel kitchen. When there's moonlight, you'll get a free nightlight in the deal.

## Installation

You need a jigsaw, a sheetrock saw, screwdrivers, and basic tools. You also need the requisite nerve to saw a hole through your roof; if you don't know what you're doing, you can hire experts to install solar tubes. Chin up, however — solar tubes come with very good installation instructions and the appropriate sealing materials. The manufacturers rely on do-it-yourselfers, so they bend over backward telling you how to do it (as well as how not do it, which is just as important).

Here are some tips for handling some of the installation challenges:

- ✔ Don't try to buy separate parts; get a complete kit, with however many extension tubes you think you need. You may want to get an extra and leave it in the box for return. Otherwise, when you find that you don't have enough extension tube, you'll probably have holes cut in ceilings and roofs, and you won't want to stop everything and go to a store. And Murphy's Law says that when you get there, they'll have just run out.

- ✔ The hardest part of the job is in the attic space, so plan your route up, then your route over to where the work is going to be done, and then how you'll sit and stand when you get over there. Also, realize that you'll be toting tools with you; the best bet is a tool belt, but if you don't have one, use a sturdy bag. Of course, Murphy says that once you reach the work location, you'll discover that you didn't remember a particular tool!

- ✔ You may want to look at the installation instructions before you buy a unit for a list of tools you'll need. If you're going to have to buy one, or rent one, add that to the cost of the project.

- ✔ Cutting through a typical roof takes more than a toy saw, so use a good jigsaw with a sharp blade. Composite shingles eat jigsaw blades, so get extras.

- ✔ Take extensive precautions to seal against the weather. (Use a good silicon sealant, although most kits will come with appropriate sealant.) If the elements can possibly get in, they will — maybe not this year, but Mother Nature has infinite patience.

- ✔ You can botch a few things on this job and nobody will ever notice, but you need to get the hole in your ceiling right. Measure twice; cut once.

- ✔ The key to a successful project is to locate the hole properly in your ceiling before you start cutting. Installation instructions are very explicit for this step, so follow them closely.

- ✔ You're going to want to make sure that you have enough extension tube. (Most solar tubes come with 4 feet of extension tube, with 2-foot extensions as options.) Get more than you think you need, save the receipt, and take back the extra. The last thing you want is to get halfway through the job, with holes in your ceiling and roof and discover you don't have enough extension tube.

## Using tabletop solar fans

Simply take a small 12VDC room fan and a suitable PV panel and wire them directly together. The hotter the sun, the more the air moves — no need for batteries. At $150, a solar fan isn't a cheap option, but it may be reasonable if your energy costs are high enough. Plus, you can use one anywhere; you don't need a plug.

These fans work nicely on porches, where you can position the panel right outside to catch the most sunlight. They're great for RV or even tent camping when the weather's hot enough. Solar fans are perfect for a pleasure boat in the hot sun. Pool houses, as well as remote casitas (small guest houses), likewise benefit.

# Cooking with the Sun

A solar oven can bake at temperatures above 350°F. You can't rely on precision temperature control, so you have to stick with foods that aren't fussy. But plenty of foods fit the bill.

You can set a solar oven anywhere you want — in the backyard, next to the barbecue, or on a balcony with a lot of sunlight. After working with your oven for a few weeks, you should be able to cook just as well as you do indoors. You just need to pick up a few tricks first. In this section, I explain these tricks — as well as how to build your own solar oven.

Solar ovens are not toy ovens, not by a long stretch. A solar oven can burn you every bit as badly as a conventional oven. You generally want to keep kids away (although a weak solar oven may be a good way for them to discover some fundamentals without much danger — a weak oven is one used in cloudy conditions or one with poor insulation or no reflectors).

## Making a solar oven

You can make a good solar oven for under $40, and it works even if you do a messy job of construction. In fact, they're so cheap, building a sloppy test oven to learn the ropes is a good strategy. Then you can build yourself a quality unit that's more convenient and lasts a long time.

Here's a parts list for the oven itself (the next section mentions pans you use when doing the actual cooking):

- ✔ Plain old cardboard box, around 20" x 20" x 18" deep; double-walled corrugated cardboard walls are best.
- ✔ A sturdy piece of flat cardboard that matches the top of your oven; if the oven is 20" x 20", that's the size lid you need, with a little overlap.

✔ Tape, with good old duct tape working just fine. Masking tape also works, but not the kind for painting because the adhesive is too weak.

✔ Standard household insulation (not white styrofoam, but hardboard style), 1" thick; this stuff is around $11 for a 4' x 8' piece.

✔ Aluminum foil, heavy duty with one shiny side at least, about 10 square feet.

✔ White glue.

✔ Flat, black spray paint designed for barbecue pits or woodstoves.

✔ Turkey bags or big roasting bags.

Figure 9-3 shows how to assemble your oven:

1. **Bend the flaps of the cardboard box out and down, and tape them down at the corners.**

2. **Prepare the insulation and aluminum foil and glue them in the box.**

   Cut the insulation to size to fit the bottom of the box, then the front and back, left and right sides of the cardboard box. For even better results, use two layers of insulation. Glue aluminum foil on one side of each piece of insulation, and spray paint the foil black. Glue the pieces of insulation into the box with the black facing the inside of the oven. If you're using fiberglass-based insulation, use gloves and eyewear; the insulation should indicate when these precautions are necessary.

3. **Prepare the lid.**

   Make sure that the cardboard lid is a little larger than the top of your oven. Then cut an opening in the lid to match the interior size of the oven. Cut the roasting bag and tape it over the opening in the lid.

4. **Poke the barbecue thermometer through the front of the oven.**

## Cooking with a solar oven

The first time, just set the oven into direct sunlight and watch the temperature climb. Play around and see how the temperature varies with different conditions (orientation to the sun, time in the sun, and so on).

Ovens that get the hottest are the ones that get the most sunlight and have the best insulation.

You can use your solar oven in the winter, but it won't get as hot because the outside air is cooler, plus you get less sunlight in the winter.

**Step 1**

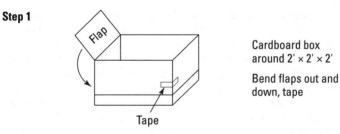

Cardboard box
around 2' × 2' × 2'

Bend flaps out and
down, tape

**Step 2**

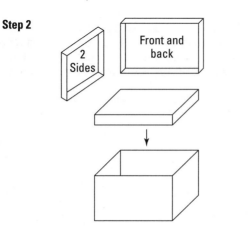

A. Cut bottom, front, back,
2 sides out of insulation

B. Glue aluminum foil on
one side of each
insulation

C. Spray paint foil black

D. Glue into box with
black inside

**Step 3**

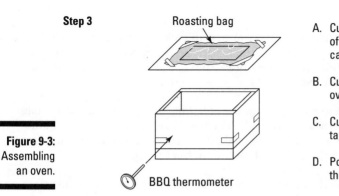

A. Cut a lid larger than top
of oven out of sturdy
cardboard, or plywood

B. Cut opening to match
oven interior size

C. Cut roasting bag and
tape over opening in lid

D. Poke BBQ thermometer
through front of oven

**Figure 9-3:**
Assembling
an oven.

To do the actual cooking, you need a good drip pan to facilitate both cooking and cleaning. Use a shallow cookie sheet or pan, shown in Figure 9-4, that fits inside your oven (because of the added insulation, its length and width need to be at least 2" smaller than the bottom of your cardboard box). You also need a black, aluminum baking pot with a snug lid — everything that's cooked will go into this pot, so this will limit your food options. You can make an oven as big as you want, however, so that you're not really limited. Some people have cooked Thanksgiving turkeys in a solar oven, to very good effect. The pot will get hotter than the temperature of the oven cavity itself (unlike a conventional oven), so be extra careful. You'll be pulling the heated pot up and out of the oven, so get handles that facilitate this upward motion. Certain pots work better than others, so try some different candidates and see what happens.

Solar ovens can get real hot, over 400°F. Not only can you get burned, but if you put napkins or other flammables onto them, you can start a fire. (The cardboard box is well insulated from the heat, so it won't catch on fire.) Treat a solar oven with the same level of respect that you treat your conventional oven.

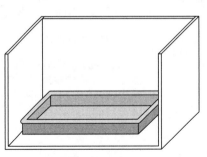

**Figure 9-4:**
A shallow cookie sheet on the floor of your solar oven functions as a drip pan.

Set a shallow cookie sheet, or similar, on floor of oven

You can use a simple oven as a warmer for rolls and potatoes and food that's already been cooked. This method is especially convenient next to a barbecue pit. As you finish burgers or brats (not the kid kind, the pork kind), put them into the solar oven — they stay nice and hot.

You may be able to cook corn on the cob. Put each ear into a roaster bag with some butter, seal it, and wait. The precise amount of time isn't important, and neither is the temperature. You can cook half a dozen ears next to your barbecue, and they taste better than the boiled variety.

Likewise, you can heat pots of beans, chili, and other canned goods such as vegetables and soups, depending on the sun.

Solar ovens aren't ideal for foods with specific times and temperatures. For example, a cake may not be a good idea.

Your solar oven already has a built-in timer. In the morning, just put something in and then aim the oven at whatever point in the sky you want it to cook. For example, aim where the sun is at 5 p.m., and when you get home from work, your dinner's ready!

## *Improving your next solar oven design*

After you know how to build a basic solar oven (see the earlier "Making a solar oven" section), you're free to play around with the design. Here are some ideas:

- **Reflectors:** You can boost the heat quite a bit with a *reflector,* which functions to increase the amount of total sunlight radiation that gets into the oven cavity. Cut some cardboard, glue aluminum foil to each piece (shiny side out), and tape the reflectors together and onto the oven (see Figure 9-5).

- **Mounting:** The best way to mount a solar oven is on a cart with wheels. That way, you can easily twist it around and point the oven toward direct sunlight.

- **Glass window:** Go to a window shop and get a piece of glass that looks around the size of a good solar oven lid. The next time you build an oven, use a box enclosure that matches. Glass shops often give out frame assemblies for free. Glass is a much better insulator than a plastic bag. Double-pane glass is even better, as long as the seal between the panes isn't broken, in which case it's useless.

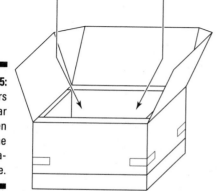

**Figure 9-5:** Reflectors for a solar oven increase the temperature.

A. Cut reflectors as shown (cardboard)

B. Glue aluminum foil to each piece

C. Tape reflectors together and onto oven

Be careful if you have an aluminum frame, which can get very hot. Wooden frames have such good insulation that you don't need a hot pad to move them.

✔ **Size:** Bigger capacity ovens with large reflectors are capable of getting really hot, and they're easier to work the food in and out of. With a good-sized, glass-windowed oven and reflectors, you may be able to roast a turkey on Thanksgiving Day, depending on the weather.

✔ **Plywood box:** The sturdiest boxes are made of plywood. You can buy pieces cut from lumber yards or get wood from a cabinet shop. The wood is square and even, which affords integrity with nothing more than wood glue to hold it together. Some of the best homemade solar ovens are made of plywood frames, which not only can you twist to follow the sun, but you can also rotate up and down to maximize the sunlight.

# Chapter 10

# Heating and Purifying Water

*T*he sun is a perfect water heater — all you have to do is expose water to sunlight, and it warms up with zero pollution. Heating water is easier than creating electricity for the do-it-yourselfer, and it usually makes more economical sense as a first step into the wide world of solar power. Even if you decide to put in a large-scale electrical system, you should first consider installing a solar water heater because you can then specify a smaller electrical system. The ratio of dollars-saved on your power bill to dollars-invested is usually better for solar water heating than solar electrical generation. You can spend less than a few hundred dollars on the simplest system, or you can spend upward of $5,000 for the most complex.

In this chapter, I go through the basics of heating water with the sun, from collectors to flow systems to mounting the systems. I also help you decide whether to do the work yourself or hire a contractor. Finally, I offer you an easy water purification project so you can break into the water distillation racket. (If you want info on installing a solar water heater supplement in your home, flip to Chapter 12. If you're interested in heating your swimming pool with a solar heater, go to Chapter 11.)

# Exploring Water Heating Systems

The most basic classification of solar water heaters is whether they're active or passive. In *active* (or *forced circulation*) *systems,* pumps move the fluids. These systems require both a power source and a controller, which decides when to pump or not to pump. *Passive systems* use no pumps to move the water. (However, this definition comes with an asterisk. When you use the pressure from your domestic water supply to drive your system, you still have a pump somewhere — it just belongs to your municipality! Or if you're on a well, the pump is deep within the Earth.) *Passive* means no pumps or moving parts are in the solar system itself. Passive systems are cheaper because they're simpler and have fewer parts. But they're also less versatile. Active systems, on the other hand, are capable of outputting more energy and working under a wider range of weather conditions. I explain a number of systems in the following sections, both active and passive.

In addition to choosing between active and passive systems, you have a couple of other decisions to make:

- *Direct systems* heat the water right in the collector. *Open-loop systems* are all direct. Water is fed into the loop and taken out of the loop after it's heated.

- *Indirect systems* use heat exchangers. A secondary fluid (water, glycol, or antifreeze) collects the heat, which is then transferred to the water via a heat exchanger. These systems are used in cold climates, where water would freeze if it were exposed to the elements. *Closed-loop systems* have isolated circuits with a constantly recirculating fluid. They're all indirect and active (they require pumps). They require entirely different engineering than open-loops and are generally much more expensive because of their increased parts count and complexity.

# Getting the Scoop on Solar Components

Every passive solar water heating system has several basic functions and components:

- **Collectors:** Sunlight must be collected and transformed into usable heat.

- **Flow:** A flow system channels the heated water to where you plan to use it.

- **Controller:** A controller makes judicious decisions on when and how to move the water, or antifreeze liquids.

- **Mounting:** You mount the collector to optimize the amount of sunlight received (see Chapter 5 for more on your solar potential).

Active systems add a pump and an active (electrical) means of controlling the pump.

In systems with copper and metal parts, using softened water is essential because hard water will calcify and corrode some pipes. You may need to find out whether your water is soft enough. If not, you may need a water softener.

## The big chill: Avoiding freeze damage

Freeze damage is a major concern. If you have no danger of freezing, you can use any type of system you want. If your climate freezes a lot, you're limited, although your options are still good. Here's how to deal with the risks:

✔ In a process called *recirculation*, some systems (active) turn the pump on when the temperature gets low enough. Moving liquid will not freeze nearly as easily as stationary liquid. This method works well, but it's inefficient; the basic goal of a solar system is to collect energy, not use it to preserve the system.

✔ Drain valves, either manual or automatic, may purge the collector and exposed pipes of all fluids. This works, but in this context, *all* means *all* because if any trace amounts linger, freeze damage can still occur. This works well, but once again, the process takes energy, and this implies inefficiency.

✔ Closed-loop systems use antifreeze and a heat exchanger. Water in the system can ever freeze. These are the most common types of systems installed on houses in North America because they're the most versatile and reliable.

Any solar water heating system presents a danger of scalding. Water over a temperature of 160°F can burn you badly enough to require medical attention. You need to understand exactly what your system is doing and where the dangers lurk. If you're going to install any kind of system, even if you don't do it yourself, you should understand what's going on inside and why. If you have children, consider all the ways they may be able to hurt themselves. When in doubt, back off, think things through, and proceed with caution. Well-designed systems account for these dangers, and the systems are perfectly safe. County codes all require the use of a *temper valve,* which mixes hot water with supply water to ensure that water temperatures that reach a user (faucet) are safe.

## *Collecting the sunlight*

Some systems are nothing more than solar collectors. For example, when covering your swimming pool with a solar blanket, you simply float it directly over the water. When you want to swim, you remove the blanket. You can do this by hand or use mechanized rollers that automatically retract the blanket for you. In either case, *you* are the controller because you're deciding when to place or remove the blanket.

As for more technical systems, a wide range of different collector constructions are available, with varying performance characteristics that depend on the application and quality of materials. Regardless of the type, some fundamentals apply.

The *efficiency* of a collector is a measure of how well it converts radiation into usable heat. Efficiency is a function of how hot the collector is; the hotter an object, the more heat it emits. The idea is to collect the heat, not emit it. In very cold weather, collectors may emit as much heat as they collect.

Running higher quantities of fluids through a collector (or running the fluids faster) doesn't result in any more heat collection because that's determined by the available radiation. However, running more fluids through usually keeps the collector cooler, which means less heat loss to the external ambient, and therefore better efficiency.

The lifetime of a collector is important in your decision-making. Some don't last more than a couple of years with full exposure to the sun. Others are guaranteed for more than ten years. If you're placing your collector in an easy-access location, a limited lifetime — and its associated cheap cost — may be best. But if you don't want to change equipment (if it's high up on a roof), expect to pay more for your collectors.

### Integral collector storage (ICS) collectors for pools

Direct, *integral collector storage (ICS) collectors* — such as swimming pool collectors and the batch collectors commonly used in Third World countries — are filled with the water that's being heated. The simplest and cheapest collector on the market is a flexible, plastic molded grid suitable for swimming pool applications. These come in 2' x 16' sheets (among other sizes) that you can lay right on the ground, or on your roof. The flow diagram in Figure 10-1 shows a very simple swimming pool collector.

Lay these collectors on the ground, connect them right into your existing pool pump system with flexible hose, and you're ready to go. Or you can lay them on a roof with a few mounting straps to keep them from sliding. If you have a Spanish tile roof, you can get collectors that curve right around the profile — it's expensive but doable.

ICS collectors are available in most swimming pool supply stores. Expect to spend around $230 for 80 square feet for the simplest models. For the kind you mount onto your roof, the added integrity (lifetime) and performance costs more.

Efficiency is very good because these collectors don't get too hot unless you're pumping the water through the collector slowly. Even then, inefficiency doesn't cost you much. (ICS collectors are notoriously inefficient in wind and rain.)

Most ICS collectors are made of copolymer plastic. Some designs are more efficient at collecting heat because they swirl the water around to maximize contact with the plastic walls. However, they also make the pump work harder, so there is a tradeoff. If your power rates are high, the extra expense of loading your pump is probably not worth the increase in efficiency.

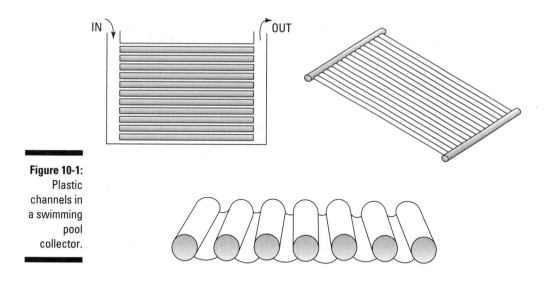

**Figure 10-1:**
Plastic
channels in
a swimming
pool
collector.

Here are a few pointers to keep in mind:

- ✔ **Always consider repairs.** Find out whether a repair kit comes with the collector or if you can get one. Ask what happens if the collector gets a hole. Look for a warranty.

- ✔ **If you stomp on ICS collectors, they can rupture.** Children have also been known to find innovative ways to destroy them. (Of course, as any parent knows, children find innovative ways to destroy just about everything!)

- ✔ **Hard or mineral-laden water can build up and block the flow as well as prevent radiation from being converted into heat.** Mineral buildup also makes the heat transfer between the plastic and the water less efficient. You can't clean out the deposits once they build up in the collector. You can get chemicals from a swimming pool supply store to remove minerals from the water, and you can also get a filter that you use on the input from your domestic water supply. (The flow rate is cut considerably, so expect a lot longer times to fill your pool.) On the plus side, the plastic materials are impervious to the caustic nature of chlorine.

- ✔ **Plastic collectors expand and contract with temperature.** Therefore, mounting must always allow for flexibility.

### Batch collectors

A version of direct ICS (see the preceding section), _batch collectors_ can be very simple: Simply put a tank of water in the sun and let it heat up (see Figure 10-2). If you paint the tank black, it absorbs more heat. If you use some reflective material such as aluminum foil to focus even more sunlight onto the tank, it works even better. If you seal it in an enclosed container with a glazed window, you get the best results.

The reflector serves to heat all the surfaces of the tank, top and bottom, as well as collect sunlight from a larger area than just the tank itself. If the reflector is big enough, you can collect a whole lot of heat. The window insulates the tank from the weather outside, preventing heat loss at night and in cold climates. (Note that you don't necessarily need the reflector inside the *glaze,* or window. (Most people think of a window as being glazed, but in the world of solar, "glazing" and "windows" are often equivalent and exchangeable terms.) Insulation around the walls of the housing reduces heat loss.

Batch collectors work well without pumps and control systems, so they're good candidates in remote locations. And they're simple to build: You can make one out of a 55-gallon drum painted black, and in some parts of the world, this is standard operating procedure.

Batch collectors can weigh a lot, more than 1,000 pounds when they're full of water, so you need to make sure that the mounting location can handle this kind of weight. Don't get fooled by their empty weight.

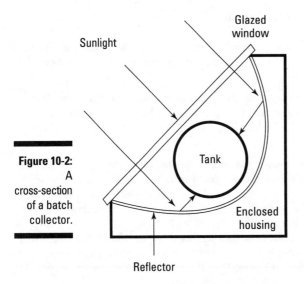

**Figure 10-2:** A cross-section of a batch collector.

### Flat-plate solar collectors

Figure 10-3 shows a very common *flat-plate solar collector.* A rectangular box (2' or 4' wide by 4' to 12' long and 8" thick) contains the parts. A copper or aluminum absorber plate at the bottom of the box is finished black to absorb the maximum amount of sunlight. Rows of fluid circulation tubes (copper, or another good heat conductor) are in direct contact with the absorber plate; as sunlight warms the plate, heat transfers to the circulation tubes and into the fluid. The absorber plate is insulated from the housing, and a glass or plastic cover (called a *glaze*) seals the unit and allows maximum sunlight in with minimum heat loss. Quality units feature a glazed window made of low-iron silicon glass.

Flat-plate solar collectors are the most widely used type of collector. They heat water very efficiently, and they have no moving parts or maintenance requirements, aside from keeping the window clean of debris (but because it takes an awful lot of debris to appreciably affect the efficiency, you don't usually need to worry about keeping the glaze clean). They work well in wind and rain, shed snow very well because of their inherent warmth, and can endure brief freezing conditions without damage. And if you combine this configuration with an ICS type collector (see the earlier section) by using large-diameter tubes to hold the water, cold damage takes even longer to occur. You can configure a multiple of these units in either series or parallel arrangements and double or triple the capacity. This is often a good tactic when you need to distribute the weight over a larger surface area of your roof.

Here are a few safety and maintenance issues to consider:

*WARNING!*

✔ Exercise care when working around any glazed collector. Not only can the parts get very hot, but the fluid is designed to get very hot as well.

✔ Depending on how large the unit is and how much metal mass is used, a flat-plate solar collector may weigh quite a bit. Carefully think out positioning onto the roof and mounting beforehand. Mounting usually means getting plenty of help.

*REMEMBER*

✔ Always purchase the mounting hardware with the collector to eliminate the possibility of dissimilar metals and galvanic reactions. (If you put certain metals in contact with each other, they corrode due to chemical reactions between the metals.) When you buy mounting hardware as part of your system, the manufacturers have already selected the proper materials for you.

✔ Glazes are fragile; they can literally shatter with the slightest provocation. Work carefully.

**Figure 10-3:** Detail on a flat-plate solar collector.

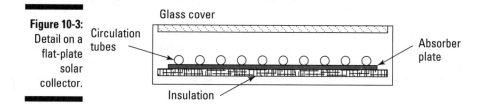

Glass cover

Circulation tubes

Absorber plate

Insulation

### Evacuated tube collectors

Figure 10-4 shows the construction of an *evacuated tube solar collector*. A row of glass envelopes (tubes) have had the air removed, which creates a vacuum and makes for excellent insulation. Copper rods inside the tubes are physically connected to a massive copper tube inside the enclosed header, through which water (or other liquid) flows. The copper rods get very hot on a sunny day, and the heat moves directly into the copper mass in the header, and from there into the fluid. When reflectors are located behind or around the tubes, these collectors are even more effective.

These collectors are impervious to freezing conditions and work very well, even when the air temperatures go below zero. They are more expensive, however. That being said, prices are coming down because so many are being built that economies of scale are kicking into high gear.

People can also make these systems with fluid-filled tubes inside, instead of rods. Then you have a direct, batch-type evacuated collector (see the preceding section). Look for more of these in the future.

Evacuated tube collectors work well in windy and wet conditions that would normally cool down other types of collectors. This type of system doesn't freeze and is extremely well insulated, yielding effective heat storage. In fact, the rods can reach temperatures exceeding 250°F while the glass envelopes are cool to the touch. But they don't shed snow, and they're way too fragile to try to brush the snow away or scrape ice off.

These collectors perform better than other types of collectors on cloudy days, and on sunny days, they work for more of the daylight hours because the tubes are always perpendicular to the sun.

Evacuated tube collectors are fragile; if the seal breaks, performance is poor because the vacuum is compromised, and insulation is poor. On the positive side, it's very easy to change a single element if it breaks. With most collectors, breakage means removing and changing the entire collector.

## Pipe dreams: Setting up the flow of traffic

Pipes and/or tubing runs from the collector to the working apparatus — usually back and forth so you have parallel sets, one pipe always hotter than the other. (One pipe is directing water into the collector; the other has the warmed water going back into the workable apparatus.)

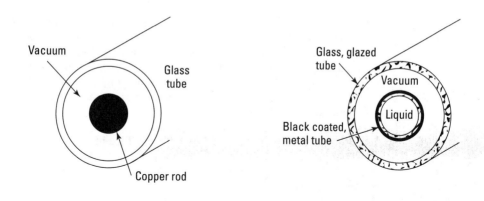

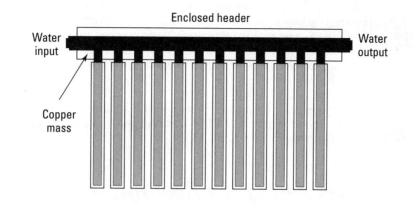

**Figure 10-4:**
Construction
details of an
evacuated
tube
collector.

Here are some tips on choosing your materials:

- **Pipe makeup:** For household systems, copper pipe is the best choice. It's flexible, easy to work with, operates over a wide range of temperatures, resists corrosion, and is commonly available. But it's heavy, especially when filled with water, so you need to take care mounting it properly. For swimming pool systems, PVC pipe is the overwhelming choice, although flexible hoses are often used with portable systems.

- **Pipe size:** Consider your goals when choosing a pipe thickness (the outer diameter of the pipe). A very thick pipe weighs much more than a very thin one when filled with water. On the other hand, a thick pipe makes your system more efficient because the pump doesn't have to work as hard.

- **Insulation:** Insulation is a must outdoors and highly recommended inside. Fiberglass pipe insulation is very good indoors, but it's useless outdoors because it absorbs water, causing it to saturate and rot. HT/Armaflex is your best bet. It lasts a long time, indoors and out. Put a plastic jacket over the insulation to shield it from the elements.

✓ **Flanges:** You need to use suitable flanges to keep the pipe in place when feeding through roofs or walls. You don't want to introduce leaks into your house when you install solar hot water heating equipment. You also need to be mindful of weight loads because flanges often bear the brunt of the forces. Flanges come with weight-load specifications, but these numbers are dependent on proper installation, so read the instructions carefully.

Make sure that you map out the flow of fluid before installation. Minimizing the length of runs is desirable for cost and efficiency. Also, bury pipes whenever possible, which prevents freezing, makes for very good insulation, and hides them from view.

Be mindful of the pressures that can build up at the bottom of the pipes if you're running them up to a second- or third-story roof (see the next section for info on pressure relief valves). You also need to understand how to purge air from the pipes when you fire up a system. (You can do so in a number of ways, and your system instructions will detail the method that applies to your particular system.)

Painting pipes the same color as your house's exterior is common, so plan on using a good primer on PVC or metal pipes. If you use insulation, it may not take paint very well.

## Completing the flow system with valves and monitors

The big functional items are in place. To keep the pressures and temperatures where they should be, a number of different valves are commonly used. And to be able to see how your system is operating, monitors are installed in appropriate locations.

✓ **Thermometers, flow meters, and pressure gauges:** These devices let you see the system parameters; otherwise, you have no way to tell whether anything's happening, let alone happening the right way.

✓ **Check valves:** These valves allow fluid to flow in only one direction. The most common are *swing valves,* which have a little gate that swings like a dog door that's constrained to only one direction. The best ones are made of bronze, and they're very reliable. Because the gate is gravity forced, swing valves only work when they're horizontal or tilted upward.

Don't use spring check valves; they load pumps too much, which constricts flow.

✔ **Pipe unions:** These connectors join two pipes without solder or permanent connection. Pipe unions are commonly used to interface with all collectors, so you can remove the collector from the system for servicing or replacement without having to cut pipe or melt solder.

✔ **Drain valves:** Drain valves are part of every solar water system because they are the means to drain the fluid out of the system (maintenance, weather safety, and so on). *Ball valves* are most reliable; they use a spinning ball with a big hole, constrained in a socket. They're either on or off.

*Gate valves* (such as the faucets used on your backyard hoses) have a twist handle that regulates the amount of flow by means of a gate lowered into the fluid path. People don't use them much on solar heating systems because you rarely want a valve that's partially open, and when gate valves are fully open, they still impede the flow of water to some extent.

✔ **Relief valves:** These valves protect against pressure buildup (on the top of your water heater you'll likely see one). They're required for every closed loop system (see the earlier "Exploring Water Heating Systems" section).

Be careful; relief valves can pop open at any time, and they spray the area with whatever's in the pipes or tank they're protecting. If they're on a hot water tank, it's going to be hot water. If they're in a solar system, the hot water may be super heated. In general, always use a drainage tube with a relief valve so that if the valve does open, you channel the flow to a safe location, such as under an enclosed porch.

✔ **Tempering valves (mixing valves):** These valves have three ports: cold in, hot in, and mixed output. A twist handle controls the output temperature at the mixed output, which flows into your home's faucets.

Tempering valves provide very important anti-scald protection and are always part of solar water heating systems. Don't be cheap here!

✔ **Motorized valves:** These valves are controlled electrically. They can be either off-on or *proportional* (a mixture of off and on, or an ability to control the amount of flow). The type chosen is a function of the application. For example, in most system applications, off/on is all that is required, and these valves are cheaper and easier to operate.

✔ **Vacuum breakers:** Vacuum breakers allow air into a system when it's depressurized. You use them to quickly drain systems, such as rooftop pool collectors. If you need to drain your system for winter use, you usually need a vacuum breaker; if you need to drain *quickly,* a vacuum breaker is definitely necessary.

## Forcing the issue with pumps and thermosiphons

*Active systems* always use a powered pump. For a swimming pool, it's the filter pump. For a closed-loop system, you need an electric pump in the closed loop. A temperature probe, or probes, and a control circuit turn the pump off and on appropriately. There are a thousand different types of pumps, and the best way to choose a pump is to buy a complete system. It's a difficult engineering problem to choose the optimum pump, so if you need to do it, call a pump store and talk to a clerk for advice.

If you have a passive system connected to your water source, the municipality is your pump. For well users, the pump is deep at the bottom of your well. In either case, you don't need to provide the pump, but technically speaking you still have one in the system.

You can get some circulation without using a pump. A *passive thermosiphon* is a device set up with a tank mounted above the thermal collector. Hot water weighs less than cold, so as the water in the collector heats up, hot water rises into the upper tank while the heavier, cold water migrates down into the collectors. Thermosiphons are entirely passive, and they are commonly used in isolated locations where electricity is not available. With no moving parts, thermosiphons are extremely reliable and efficient.

## Directing traffic with controllers

A solar blanket for your swimming pool is controlled by applying the blanket or retracting it. You're the controller. Other solar water heaters, however, require temperature controllers that are a little more high-tech. Take a look:

- ✔ A collector connected to your pool pump works whenever the pump is working. The controllers are electric clocks that cycle every 24 hours; you adjust the mechanical position of *off* and *on* tabs.

- ✔ Domestic water heater systems often use an electronic controller that takes temperatures, time of day, and other factors into account. These controllers are usually digital, and you can easily program them for your particular application. They also prevent scalding fluids, and they control valves to evacuate systems in freezing conditions.

- ✔ *Differential temperature controllers* have two separate temperature probes located in various parts of a system; they open and close valves in response to programmed logic.

# Addressing mounting concerns

If you mount solar panels on your roof, you can accomplish several things:

- ✔ Maximize your solar exposure, so you don't have to worry about shading from your house, trees, and the like.

- ✔ Use the roof's pitch to angle your solar collector to intercept the most sunlight without the need for complex brackets.

- ✔ Shield a portion of your roof from direct sunlight, moving that heat to a different location. If you were to cover your entire roof with solar panels, no direct sunlight would ever hit your roof. Your attic would be cooler, as would your entire house!

 When you're deciding whether to mount equipment on your roof, consider how long it's going to be before you need to put up a new roof. Having the roofers contend with solar collectors makes the job more expensive, so you may want to opt for ground-mounting or change the roof first. Here are some other factors to consider:

- ✔ **Weather:** Ultimately, understanding when and how to maintain your system so it won't be destroyed in bad weather is your responsibility. Most solar collectors include warranty restrictions against freezing damage, so it'll be on your dime if you don't follow the rules. If you need to follow certain restrictions against freezing, how you mount the collector may be important. For example, if you need special valves to drain your system, you need to make sure that the valves are readily accessible.

  The wind can exert big forces on your panels. If they aren't mounted properly, you can end up with a destructive flying object (DFO) in the neighborhood!

- ✔ **Proximity to the domestic water heater tank:** Close is best. The farther from the storage tank you place the collector, the more inefficient the system will be. In the winter, heat loss from uninsulated pipes may actually offset heat gains.

- ✔ **Damage from wildlife:** Animals can destroy panels. Do you have raccoons? How about birds that may consider a nice solar panel to be the perfect toilet? Cows can destroy a panel in three seconds flat. Goats may prefer them for dessert. The list goes on.

- ✔ **Human damage and safety:** Consider whether the panel location lends itself to vandalism, theft, or target practice. If your kids are likely to see your solar collector as a state-of-the-art set of monkey bars, you may want to put the panels up and away.

- ✔ **Potential failures and drainage issues:** A small hole in a closed-loop circulating system can turn into big trouble. If you put a solar collector below the level of your swimming pool, a small hole can literally drain your entire pool.

- ✔ **Available space:** Redundancy may be desirable — instead of using one large collector, you can use two that are half the size.

- ✔ **Available sunlight and desire for efficiency.** With adjustable mounting brackets, manual adjustment allows for varying the elevation angle over the course of the year. The winter sun is much lower in the sky than in summer; by adjusting the angle of your collector, you can optimize efficiency. You may also want to collect more sun during the morning versus the afternoon; in this case, you adjust the east/west orientation. (For more on maximizing solar potential, see Chapter 5.)

- ✔ **Appearance:** Passive solar panels aren't beautiful objects of art. If you want to put collectors on your roof, go up there and lay a dark-colored tarp down and then step back and see what it's going to look like from the street. Does it destroy the aesthetic harmony of your roof? Roof lines can drastically influence the appearance of your house.

  How does your solar collector affect the neighbor's view? You may not want to make enemies out of your neighbors by shifting the "ugly burden" over to them.

# Deciding Whether to Do It Yourself

You have the option to install solar systems entirely on your own, or you can hire a contractor to do the entire job or any portion of it.

## Weighing your options

You can design a system reasonably well without professional guidance. Still, climbing around on a roof and soldering pipes isn't for everyone. In addition to your own skills, here are some factors to consider when deciding whether to install a solar water system on your own:

- ✔ **Cost:** You can save a lot of money by doing your own installation. First, the contractors mark up the parts, so you can get them cheaper if you look around. And of course, contractors have high labor rates, which include insurance, licensing, and so on.

- ✔ **System problems and installation risks:** The vast majority of problems with solar water heating systems occur from faulty installation. Even licensed contractors make mistakes, but at least if you hire a contractor, you'll have recourse.

If you break a big, heavy collector, you have a massive object with a voided warranty that you need to truck back to the factory. Is it worth taking this risk? And mounting on your roof requires holes in your roofing material. If you don't do it right, you can create leaks. Nothing is worse than finishing a nice new project, only to find that you have an even bigger project fixing the leaks in your roof during a heavy rainstorm!

✔ **Safety risks:** Getting the panels onto your roof may be the most difficult part of an installation. Use common sense about whether to hire somebody to mount your collectors! Consider the amount of damage that may occur if you fall and weigh it against the potential cost savings of doing-it-yourself. It's quite a mismatch — you can't argue that.

Thousands of accidents occur each year when people fall off roofs, and the injuries can be serious, maybe even fatal. Composition shingles can be slippery when they're old because the surface grit comes loose. This can catch you by surprise (and the only surprise you want is on your birthday). Roofs are always dangerous when they're wet. Roofs with steep pitches are always dangerous, period.

✔ **Warranties and insurance:** Sometimes when you do your own installation, the manufacturers' warranties don't apply. Check first with the manufacturers. (Some manufacturers refuse to sell their parts to private parties; they work only through licensed contractors.) Also check with your insurance company to see what varieties of damages will be covered by an incorrectly installed system.

✔ **Tax incentives and rebates:** Sometimes tax incentives and rebates are contingent upon a system being installed by a qualified and licensed contractor.

✔ **County codes:** A lot of locales require county building inspectors for the plumbing job. You can do your own plumbing and still pass, but you may want to consider this in your decision.

## *Going it alone: A guide to the skills you need for different projects*

Most swimming pool systems are reasonable projects for do-it-yourselfers because they use PVC, don't exposure you to high temperatures, and are mostly passive. If you make a mistake, the consequences aren't dire. You may flood your roof or your yard, but you won't destroy your home.

Most domestic water heater projects require plumbing skills, at the very least. You need to understand all the valves and system pressure requirements. As you do the installation, you need to route the pipes to the collectors and then back to the domestic water heating tank. Weight considerations are important. You need to use the right feedthroughs, flanges, and clamps. But this stuff can be easy if you do your homework. All kinds of online resources can help you plan your installation.

Copper tube soldering is tricky, so if you're hesitant about working with copper tubing, get some tubes, a blowtorch, solder, and flux, and try some solder joints before you decide to do the whole job. If you're like me, you'll completely shank the first try, the next won't be so bad, and the tenth will be very good.

If you design a system yourself, you may either overspecify or underspecify the capacity, but this is no big deal. The *block diagrams* (the system prescription, including all the parts and where they go in relation to each other, and in relation to the mounting environment) are the most important thing to get right. You have to have the valves, the pumps, and collectors, in the right location. Then you need to build the system according to plan.

You can avoid a lot of problems by ground-mounting. If you have enough room around your house, this may be the way to go. Plus you get good access for servicing and upgrades.

Before you begin the installation stage, make a detailed list of the tools you need. Do you have them? Can you rent them? Consider how you're going to get the collectors up on the roof (Two large friends? A winch? Elephants?). And as always, measure twice; cut once.

At the very least, get a quotation or two, even if you're inclined to do your own installation. You can learn a lot. And here's some of the best advice in this entire section: Try to find a house you can go to that has the same type of system you're considering. Take a detailed look at all the parts and what sort of tasks are going to be required. There's no substitute for getting your hands on the real thing.

# Purifying Your Drinking Water

Here's a great bite-sized solar project for the do-it-yourselfer. Not only can you build a solar water-purification system such as the one in Figure 10-5, but you can also design it. Designing is just as much fun as building, and it's more rewarding because the system's entirely yours (unless it doesn't work, of course — then you need to figure out how to blame somebody else).

The system uses distillation, a process that can remove salts, microorganisms, and even chemicals such as arsenic, leaving you with pure $H_2O$. Here's how it works: If you leave salty or contaminated water in an open container, the water evaporates and leaves the contaminants behind. If you heat the water, the process speeds up considerably.

After the water evaporates, the water vapor condenses on the glass window and drips down into the catch trough. Tilt the catch trough just slightly and put a bottle or other container underneath the low end, and voila! Purified water.

You can make a purification system as cheaply or as expensively as you want. People in Third World countries use large, efficient versions of this same exact device that are capable of purifying hundreds of gallons a day. A system the size of a microwave oven can yield up to 3 gallons of purified water on a sunny day. Here's what you need for a basic solar still:

- A wooden (or even better, sheet metal) enclosure as shown in Figure 10-5; if you want to get imaginative, find a good metal box and cut a hole for the glaze cover

- Reflective material such as aluminum foil (shiny side out)

- Black paint (the kind used for barbeque pits works best)

- Glass (you don't need glazed glass; just use plain old window glass if you like — you can get pieces of discarded glass from window shops for little more than a smile)

- Insulation (the white foam stuff is cheap, effective, and easy to work with)

- Glue (silicon sealant or similar weather-resistant material)

- A tray that's black or has some other quality that absorbs heat

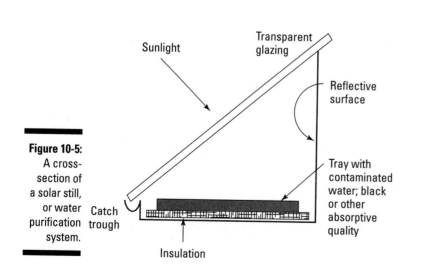

**Figure 10-5:** A cross-section of a solar still, or water purification system.

South

Sunlight

Transparent glazing

Reflective surface

Tray with contaminated water; black or other absorptive quality

Catch trough

Insulation

Now take a look at the assembly:

1. **Paint the exterior of the box black to enhance absorption.**

2. **Install the reflective surface at the back and side walls of the enclosure and glue the insulation to the bottom.**

3. **Put a tray of contaminated water inside the enclosure, place the glass on top, and aim toward the sun.**

   That's it — you're done!

The first few times you use this device, the water may taste a little odd. Let the system "sweat" for a few weeks, and the bad taste will go away.

When designing or improving your unit, make filling the inner tray with water convenient. Position it by a hose, for example. Then you don't have to carry water to your system. Configure some kind of funnel through the sidewall so you can pour the water right into the tray without spilling. Make the unit so you can remove the glazing top and clean out the tray because the contaminants are going to remain behind. You'll get a good idea of how bad your tap water is when you see what's left in the tray over a period of time!

# Chapter 11

# Diving In to Swimming Pool Projects

*A*lmost everyone wants to conserve energy and help reduce greenhouse gases. Certain energy expenditures, such as driving, heating, and lighting, are just plain necessary for most people (although you can still reduce them). Others, including swimming pool heat, are entirely optional creature comforts. If you adopt the energy conservation tenet, the only real solution is to use solar energy to heat your pool because it's the only method that's completely pollution-free.

Luckily, swimming pool projects are very accessible for the do-it-yourselfer. You can spend as little as $200 and as much as $5,000 or more. In this chapter, I show you how to choose the best system for your needs, and I also present a couple of fun projects that you can build yourself — and in the process, you can discover a lot about solar energy systems.

## Warming Your Pool With Solar Pool Covers

The cheapest and most effective solar-heating system for your swimming pool is a solar cover. You can get one for around $0.30 a square foot, so covering an average size pool costs about $130.

Using a plastic cover alone can extend your swim season a couple of months on each end of the summer. In the Midwest, for example, you may be able to swim in a completely unheated pool from mid-May until mid-September — four months. But if you use a cover, you can swim from mid-April until mid-October. And the water will be more comfortable over the entire season, which means you'll use the pool a lot more.

## Understanding how covers work

Despite the fact that the surface area of most swimming pools is large, very little of the sunlight that strikes the pool gets stored as heat. Pool water is transparent (hopefully), so it simply passes sunlight right through.

A cover absorbs the sunlight and then transfers that heat to the water. Some covers are black for this very reason, but the most widely used are made of inexpensive clear plastic that looks like the bubble wrap used for packing. These covers are made of a special material designed to store the sunlight as heat, and air bubbles work as insulation so that the heat becomes trapped in the pool water.

Covers can do the following:

- ✔ **Prevent heat from escaping your pool:** Retaining heat is especially important at night, when the air is cooler than the swimming pool. In an uncovered pool, you can easily lose 4°F or 5°F in temperature over the course of a cool night.

- ✔ **Directly convert solar radiation into usable heat:** In an in-ground pool, a cover can increase the water temperature by 5°F for each 12 hours of coverage, and it can increase the temperature by up to 10° when covered for 20 hours. When you put a pool cover over your pool, you can stick your hand down underneath it and literally feel the heat.

- ✔ **Limit evaporation:** Each gallon of evaporated 80°F water removes about 8,000 Btus (2.34 kWh) from the pool.

- ✔ **Reduce chemical depletion:** This advantage reduces cost, and you can consider it a form of energy conservation.

## Reviewing operational considerations

You do have to place covers onto the water and then remove them. I freely admit covers can be a big hassle. You may decide to increase the surface area of your solar collectors just so you don't have to deal with a cover. But just about everybody who's ever used a cover can tell you it works. And all kinds of systems are available for removing and replacing the cover — some manual, some automatic. In fact, you don't need to install any retraction system at all; you can just fold them and unfold them, as needed.

*Manual systems* cost around $300. These roll up the cover, much like a window blind. But you still have to pull it back out over the pool, which may mean you have to get into the water first. In either case, cranking the handle isn't easy; people who don't have much upper body strength may have a hard time.

*Automatic systems* can cost more than $1,000, and the installation is a bear. But all you have to do is press a switch and voila! Back and forth (at least in theory, because all mechanical systems are prone to failures).

Don't store a cover in the sunlight when it's not on the pool. They get really hot, and the plastic material degrades.

## Getting the most out of your pool cover

What's great is that you don't need to cover your entire pool to reap the benefits. Many pools are kidney shaped, but you can float a rectangular cover over only a portion of the pool, and it'll still help considerably.

Use a cover only when you know you won't be using the pool for a few days at a time. But when you swim everyday, you can leave it off. You don't need a cover during the hottest months, which is when people want to swim the most. Just fold it up and store it for the season (store it out of the sunlight).

Covers can be dangerous! If a person (particularly a child) falls onto a cover on a pool, he or she can get wrapped in the cover, which can be very difficult to get out of. Or if somebody swims under a cover and tries to surface for air, the swimmer won't find any. Be careful, and provide appropriate supervision.

Of course, you want to replace the cover when it begins to fray. When plastic bubble wrap covers get old, plastic shards can get into the filter and may even cut the filter paper, necessitating a new, expensive filter.

# Looking at the Pluses and Minuses of a Solar Pool Heater

Using a solar-heating system for your swimming pool makes a lot of sense. For starters, installation can be very easy — the simplest systems take about an hour. You can also design and build systems over a wide range of costs and complexities, zeroing in on the system that fits your precise needs.

Swimming pools already have pumps and controllers in their filtration systems, so tying in solar collector panels with existing equipment is pretty straightforward.

Swimming pool solar collectors are inexpensive because they don't need to be *glazed,* or sealed. Plus they can have a flimsier construction because failures aren't dire. You can get very large surface areas for a low price.

On the downside, solar collectors increase the workload of your pool pump, thereby drawing more electricity to filter the same amount of water. That's right — solar panels for your pool are not completely free of energy bills! You may see them advertised as entirely free, but they're not.

If your system is properly designed, using solar to heat your pool is a lot cheaper than other options such as gas heaters, electric heaters, and so on. But if your system is poorly designed, it can end up costing you more electricity than what you gain in heat. And a poorly designed system can wear out expensive pump motors really fast.

# Making Your Swimming Pool Efficient

Boosting the efficiency of your swimming pool is an important first step. Here's how to make your swimming pool energy efficient so that when you add solar, you get the most bang for your buck:

- ✔ **Reduce bends in the piping.** Sharp bends in the PVC piping slow down the flow and require more power to do the same job. Unfortunately, many pool installers completely fail this simple requirement. If your pipes are all over the place, rebuild the system — PVC is a cinch to work with.

- ✔ **Make sure that all valves are working properly.** If you have gate valves, replace them with ball valves, which are more efficient. Make sure that all ball valves are completely open or closed.

- ✔ **Keep the filter clean.** A dirty filter loads the pump, which costs a lot more power. If your filter is old, replace it. Cartridge filters are better than diatomaceous Earth.

- ✔ **Install a smaller, higher efficiency pump, and run it less each day.** Use the smallest, most efficient pump possible — ¾ horsepower is usually sufficient. If your pump is a few years old and wasn't designed with enough capacity for solar panel use, buying a new one will probably be economical. Most people will find that they can run their pool pump for much less time than they currently are doing and still achieve satisfactory cleanliness. Give it a try.

A large power pump filters your pool water faster, and some people like that because it means you can run the pump less (which means you listen to it less). But here's the problem: If you're planning on putting up solar panels, the amount of heat they put into the pool is a function of how much *time* water is flowing through them; the quantity of water isn't as important. So if you have a large pump that moves water quickly, you're not optimizing your solar panels.

✔ **Install windbreaks around your pool.** Wind can increase evaporation 300 percent or more, which wastes a lot of energy, much more than you may think. (For info on using trees and shrubs as windbreaks, see Chapter 8.)

# Understanding a Simple Starter System

Figure 11-1 shows the simplest system possible for heating your swimming pool using a solar panel. Your pool system already includes the pump, controller, and filter, along with PVC pipes that route the water flow. Simply break into the PVC line after the filter and run a couple of flex hoses (or PVC, if you prefer) to the solar collector panel, which you can lay out on the ground or set against a hill to achieve some angular tilt toward the sun.

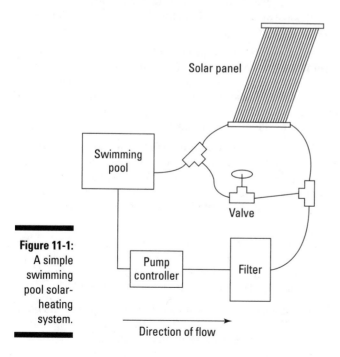

**Figure 11-1:**
A simple swimming pool solar-heating system.

When the valve is closed, water runs exclusively through the solar collector panel, heating the water. As you open the valve, water flows less and less through the collector panel because the path through the valve is so much easier. In this way, you can adjust how much heat is going into your pool. In the middle of summer, you may not want any additional heat.

Solar collector panels are available at most pool supply stores. You can get a 4' x 20' panel for around $220. Flex hoses cost around $35 apiece. Adaptors are sometimes necessary.

Here are some ways to run the system for best results:

✔ **Place the solar collector panel where it sees the most sunlight when the pool pump is running.** If you're not running the pump in the late afternoon, don't worry if the collector is shaded at that time. You may need to adjust the pump run time to match sunlight conditions.

✔ **If you place the solar collector panel on your roof, try to minimize resistance to the water flow.** Roof-mounted collectors may work better, but the piping runs may be more complex, and you're increasing the workload of your pump. If you use PVC, minimize tight angles and use 2"-diameter pipe.

✔ **Keep the solar panel out of the wind as much as possible.** Wind causes convection cooling, which results in less heating of the water flowing through the collector. However, remember that placing the panel in the sun is more important than keeping it out of the wind.

✔ **For best heating results, run the pool pump during the sunniest time of the day; running it longer will result in more heat in the pool.** However, in the middle of the summer, your pool may actually get too hot if you run your pump during sunshine hours with a solar collector in line; run the pump at night instead. You can actually cool the water in a pool by running the solar collectors at night.

✔ **Account for whether you're using a pool cover.** If you use a pool cover (see the preceding section) in conjunction with the solar panel, you can raise the water temperature up to 15°F. Without a cover, use about 50 percent more surface area in your collectors to achieve the same temperature gains (assuming, of course, that your collectors are in sunlight).

✔ **If you want to use two or more solar collector panels, connect them in parallel.** Flexible solar collector panels that are designed to fit together in a parallel "ganged" arrangement, as shown in Figure 11-2, are available. (For an explanation of series and parallel, see Chapter 4.)

Parallel-ganged solar collectors are the type you commonly see on roofs. You can put as many of them together as you want. The more you use, the easier the pump's job is and the more heat you collect.

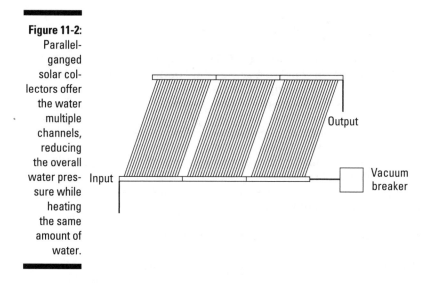

**Figure 11-2:**
Parallel-ganged solar collectors offer the water multiple channels, reducing the overall water pressure while heating the same amount of water.

Input

Output

Vacuum breaker

# Deciding on the Size of Your Collectors

The general rule is to use a total collector surface area that's about half the surface area of your pool. But this varies quite a bit. Sizing your collector surface area depends on a number of factors:

- ✔ **The pump's running time:** The longer it's on, the more heat you collect in the pool for a given collector size. So you can use less collector size if you run the pump longer (up to a point; you still need sunlight, and if you run the collectors for too long, the sunlight won't be available). Running more water faster through the collectors doesn't do much good.

- ✔ **Pool location:** If you have an aboveground pool, the heat loss is much greater. You need a collector with more surface area.

- ✔ **Solar potential when the pump is on:** Note how your solar panels are oriented. How much sunlight do you get? Many roofs aren't oriented south, so collectors mounted flush aren't working as efficiently as they could be (see Chapter 5). Hazy days don't get much radiation, either. In those cases, you need more surface area. For example, in Seattle you will need more surface area than in Phoenix.

- ✔ **Shade:** If you have shade problems, you can solve these, but you may not want to. Cutting down Old Man Oak isn't much of an option if your blood is truly green. Otherwise, choose a larger collector.

- ✔ **Wind:** If you have a lot of wind, your collectors will run inefficiently. Not so with glazed collectors, but the open ICS collectors (see Chapter 10) used for swimming pools are very sensitive to wind. So if you're in a windy climate, increase your collector size or increase the pump time per day.

# Installing a Complete System

Swimming pools cost a lot, upward of $30,000. If you don't have any kind of swimming pool heater, your useable season may be around four months of the year. If you install a swimming pool heating system, you can get eight months. This explains the popularity of pool heaters.

Figure 11-3 shows a diagram of an entire full-scale, professional-grade, swimming pool solar system. When the pump is on, the controller decides whether to activate the solar collectors by measuring the temperature at the collectors and the temperature of the pool water. When heat's available at the collector, the motor valve opens, and the pump moves water up into the collectors and back down to the swimming pool.

Here are the operational details concerning the valves:

✔ When the controller deactivates the motor valve and no longer allows water to pump into the collectors, the vacuum breaker allows the system to purge itself of liquid for two reasons:

✔ You don't want water in the collectors at night because it cools down; when you first activate the collectors the next day, you'll dump a bunch of cool water into your pool, which is at odds with the goal.

• Purging the water ensures it won't freeze up and burst the pipes or collector.

✔ The check valve prevents water from flowing backward through the pump and filter.

✔ Use the manual valves to deactivate the system and purge the pipes in off-seasons or during maintenance.

People almost always mount the solar collectors on their roofs. The collectors need to have a slight downward tilt to facilitate draining. Make sure that you don't injure the roof and cause leaks.

If you put your solar panels on your roof, keep in mind that your pump will have to force water up to the height of the solar panels. Some pumps may be unable to do this if the panels are much higher than the pool. Find out what specs apply to your pump. The appropriate specification is called *head pressure,* and it's given in terms of feet. To be safe, measure from the surface of the pool to the top of the collector and add 4'.

Piping is almost always 2" PVC. Keep 90° bends to a minimum because they increase the resistance to water flow, which makes the pump work harder.

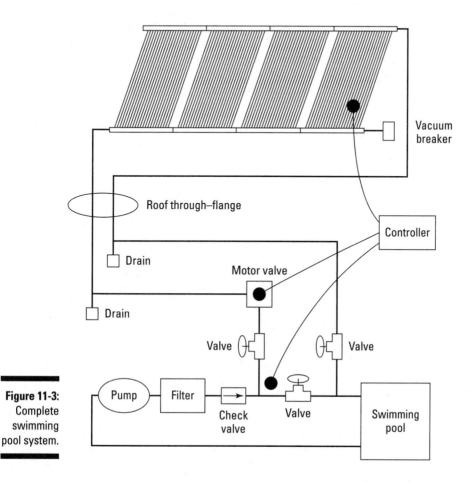

**Figure 11-3:** Complete swimming pool system.

You can design your system for future expansion so that you can change individual parts whenever you want. That way, you can experiment with operating parameters without worrying about blowing something up or causing damage to your house. The worst you can accomplish is a flooded roof or backyard — or maybe a bruised ego.

# Creating a Simple Homemade System

A simple homemade solar pool-heating system, as shown in Figure 11-4, performs ideally, and you can build it entirely on your own. You lower the submersible pump into the pool water so it'll always be primed. The pump runs off the 12VDC that the PV module supplies. The amount of water the pump passes through the solar collector panel depends entirely on how much sunlight is incident on the PV module: At night, the pump doesn't run at all. On cloudy days, the pump runs slowly. When it's sunniest, the pump is cranking full speed.

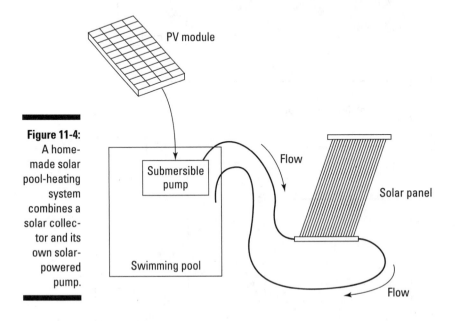

**Figure 11-4:**
A home-made solar pool-heating system combines a solar collector and its own solar-powered pump.

Because the heating system pump is separate from your filter pump, you can run your pool filter pump for only two hours a day without having any effect on how much heat you get from your collectors. In some locales, the power rates are much different during day and night; you want to run your filter pump at night, but you need to run your solar collectors during the day. With this system, you solve that problem.

A solar PV panel to power the pump costs about $130, and a pump costs around $40, for a total of $170. You keep your power bill for the pool pump to a bare minimum, perhaps saving you enough in one season to pay for all the parts of your new system.

You need to get a submersible pump with enough head pressure to raise the water up to the height of your collector. To determine ideal head pressure, measure the vertical distance from the water surface to the collector and add 4'.

If you put the collector on the ground, almost any pump will do the job. If you can put the collector below the surface of the pool, you can get help from siphon physics.

You need wiring, a solar collector, piping, and a PV module. A 12VDC panel should be strong enough to power your pump.

# Chapter 12

# Supplementing Your Domestic Water Heater

. . . . . . . . . . . . . . . . . . . . . . . . . . . . . . . . . . . . . . . . . . . . . . . . . . . . . . . . . . . .

## In This Chapter

▶ Surveying systems to supplement your domestic water heater

▶ Doing your pre-project homework

▶ Installing systems for a variety of applications

▶ Using solar-supported radiant heat floors

. . . . . . . . . . . . . . . . . . . . . . . . . . . . . . . . . . . . . . . . . . . . . . . . . . . . . . . . . . . .

*H*eating water for your home comprises anywhere from 17 to 28 percent of your power bill, and most hot water goes right down the prover-bial drain. From washing dishes and clothes, to taking baths and showers, to cooking food and warming your pool water, the range of activities that requires hot water is extensive.

An electric water heater supplying a family of four uses 11.4 barrels of oil per year. That's more oil than is consumed by a medium-sized auto driven 12,000 miles per year, at 22 mpg, which uses only 11.1 barrels. When you install a solar water heater to supplement your domestic water supply, you're pro-ducing all the energy on site, in your collectors. There's no pollution at all (aside from that used to manufacture the equipment, which is referred to as invested energy).

In this chapter, I run through some options for supplementing your domestic water heater with solar power. I discuss the various systems, explain instal-lation, and even talk a bit about warming your house with radiant heat. (For details on the basics of solar water heating, including info on solar collectors and other components, see Chapter 10.)

## Looking at the Various Systems

A wide range of supplemental water-heating systems are on the market, but only three — which I discuss in the next section — have stood the test of time. You can install any of them yourself, although you may not want to. Even if you intend to have a pro do the installation, make sure that you understand the various alternatives in order to make the best decision on which system you want. You also need to understand your system in order to get the most out of it. In this section, I explain your options and some of the factors that should go into your final selection.

## Highlighting the most popular systems

Literally dozens of different types of systems are available, but only a few have stood the test of time. Sticking with proven entities is always a good idea, despite the extravagant claims you may hear regarding new technologies. In the real world, Murphy has a voracious appetite, and until something is tested thoroughly, the jury is still out.

### The integral collector system (ICS)

ICS is the simplest and cheapest system, and it's ideal for mild climates. But it also has the most potential for freezing damage because the collector holds the water being heated. Big batch collectors, however, can withstand longer freezing conditions than less bulky systems.

These systems are passive (they don't have internal pumps) and are normally plumbed directly between the cold water supply and the water heater, making the plumbing job easy and straightforward. Whenever someone opens a tap in the house, water flows from the collector into the water heater. If the water in the collector is hot enough, the domestic water heater doesn't need to add any heat at all.

Most ICS systems need to be drained of water when in danger of freezing. The systems use manual or automatic valves. A burst collector costs a lot of money to fix. They're big and bulky, and you may end up having to send them back to the factory. But they're cheap and easy to use and install, which is why they're so common. You just need to understand the operational drawbacks if you're going to successfully utilize one.

### Drainback systems

*Drainback systems* (direct, ICS) do exactly that: They drain the fluid out of the collector and exposed pipes when no sunlight's available — at night, for example — or when there's no more need to heat the domestic water because it's already at the preset temperature. These systems use a special tank for holding the drained fluid. These types of systems are medium on the cost scale and can be used in cold climates. There is less risk of bursting pipes than a conventional ICS system, but more risk than a closed-loop antifreeze system (see the next section).

## Tapping into your hot water supply

Hot water storage is never perfect. Over time, all storage tanks lose heat, resulting in inefficiency. So the best time to use hot water is right after your storage tank has been charged up — in other words, afternoon. This is less true if your storage tank is very well insulated.

By changing some habits, you can literally reduce the energy workload on your conventional water heater to zero. For example, if each member of the family takes a shower at a different time of the day, a batch collector can heat up again between showers. If you take your shower in the afternoon, after the tank has been heating up during the midday sun, your residential heater has to do very little work. You can run your dishwasher or wash clothes in the late afternoon instead of the morning. The list goes on — use your imagination. Avoid using hot water first thing in the morning as much as possible (this may be difficult, granted).

In the winter months, when you're getting very little solar energy, the timing of when you use hot water doesn't matter as much because it almost all comes from your electric or gas heater.

### Closed-loop antifreeze systems

Closed-loop antifreeze systems use fluids other than water to collect the heat; then a heat exchanger transfers that heat into the domestic water supply. These systems are far and away the most widely distributed type systems in the world because they work in almost any climate. They're the most expensive as well.

These systems aren't entirely free of problems. The antifreeze solution breaks down when the weather gets really hot and then turns corrosive. These types of systems require service, and owners need to stay on their toes in order to prevent big problems.

## Skimming through the lesser systems

A few other systems deserve mention as well. However, these don't warrant a top recommendation because they're applicable to specific conditions. If you want to understand the technology, it's worth reviewing these systems. If you buy a house with a system already installed, you may encounter one of these systems:

 ✔ **Recirculation systems:** These systems use a temperature sensor to determine freezing dangers; in cold temperatures, a pump starts recirculating water through the exposed collector and pipes. Water in motion is extremely difficult to freeze, so the water in the system can be below freezing without actually becoming solid. This works most of the time, but not all. And when in recirculation mode, the system loses a lot of heat, which is the opposite of what you're trying to do in the first place. Avoid these systems. They don't offer enough pros to offset the loss of efficiency.

✔ **Thermosiphon systems:** Thermosiphon systems work well in theory, but you need to mount the storage tank above the collector (at a higher elevation, because hot water rises). You see these in Third World countries with moderate climates because they're very cheap and easy and reliable. Some ICS systems also use the thermosiphon effect to good advantage.

✔ **Open-loop flooded systems:** These systems are similar in nature to ICS systems, but they use a flat-plate collector and a storage tank. They don't really offer any advantages over an ICS system, and they freeze up faster and tend to clog faster as well. Avoid them.

## Considering factors that dictate the type of system

Several operational factors dictate the type of system that will work best for your particular application:

✔ **Freezing weather:** The most important factor in determining the best system for your needs is your climate — in particular, the likelihood of freezing. Water expands when it freezes, and almost everybody who lives in cold climates knows about burst pipes.

Copper is the piping of choice for domestic water heater systems because it ages well, works over a wide temperature range, is easy to work with, and withstands hard water. But a well-insulated ¾" copper pipe full of water freezes in less than five hours at 29°F. A burst pipe is a greater danger in most systems than collector damage.

If you live in freezing conditions, consider a closed-loop antifreeze system. If freezing is occasional, you can use any system you want, as long as you purge the system of water before freezing conditions occur. Design your system with appropriate drain valves, in easily accessible locations.

✔ **Potential high-temperature conditions:** Too much heat can be every bit as damaging as too much cold. ICS collectors may burst if the water inside boils. Antifreeze systems don't suffer this problem. If you have any other type of system, and the weather gets real hot, just slightly open a hot water faucet anywhere in the house, and fresh, cool water will constantly feed through your collector. Don't worry about wasting energy; you're not paying for the wasted hot water, and it's not costing any pollution. In fact, you can store hot water in your bathtub for later use, if you want. Just add more hot water at that time, if you need to.

✔ **Quantity of hot water produced:** The average household uses around 10,000 to 15,000 Btus per person per day. Collectors are rated in terms of how many Btus per day they generate, given average conditions. Realize that your hot water output varies over the seasons and weather conditions. Install a system that's smaller than you need as opposed to larger. Extra capacity is just wasted money.

✔ **Purity of your water supply:** Mineral buildup can decrease the efficiency of your system. Try an experiment: Pour water into a clean container and then let the water evaporate. Repeat this a good number of times, say 20. Is there a white residue? Any kind of residue? Now imagine exposing your solar collector to this. Over the course of a year, you build up much more than what you can see in your container.

You may want to consider a water softener. If you have one, you definitely want to run the water through the water softener before it gets to your solar system. And you may want to consider installing an even more extensive water purification system. The subject is beyond the scope of this book, but you can research the details over the Internet. Check out www.realgoods.com.

Table 12-1 shows you how the main system types compare.

| Table 12-1 | Comparing Water Heating Systems | | | |
|---|---|---|---|---|
| *Type* | *Freeze Risk* | *Heat Risk* | *Water Temperature* | *Approximate Installed Cost* |
| ICS | High | High | Inconsistent | $2,200 |
| Drainback | Low | Low | Inconsistent | $4,000 |
| Closed-loop antifreeze | Lowest | Lowest | Consistent | $5,000 |
| Recirculation | Low | Low | Inconsistent | $4,000 |
| Thermosiphon | High | High | Very inconsistent | $1,600 |

The type of domestic water heater you already have in your house can dictate what type of solar water heater you install. If your domestic water heater is nice and new, find a solar system that supplements it. If you're ready for a change of domestic heater, combine the decision on which type with the decision on a new solar system. Integrated systems work better.

# Following Good Advice (Mine)

Regardless of whether you decide to install a system yourself or have a contractor do the job, there are some definite tips you will want to keep in mind.

I highly recommend that you try to check out a real, live system of the type you decide on. Will the neighbors let you look at theirs? Ask your potential supplier for references and then find out whether you can go look at the system and ask the owners what they think. Here are some of my recommendations:

- ✔ **Opt for a complete system — it works the best.** Mismatched components reduce efficiency and may even void warranties.

  Make sure the system includes monitoring devices. Temperature gauges and flow meters indicate proper operation and let you know when the system needs servicing.

- ✔ **Seek out quality parts.** Quality parts mean fewer service calls. Reputation is critical. Look at consumer Web sites and ask around, and review the owner's manual before you buy. Does the manual make sense? Is it easy to read? Is it logical? If you get a bad feeling about the manual, cut and run.

- ✔ **Pay attention to contractor quality — it's a big factor.** Too many cheap jobs end up being expensive. The number-one failure mode of all solar water systems is faulty installation.

  Check whether the manufacturer and installer are affiliated with the Solar Energy Industries Association (SEIA). With all the buzz over solar these days, all kinds of unskilled, unsavory folks are getting into the business. Use your common sense; if it sounds too good to be true, it surely is.

- ✔ **Understand the contract terms.** Every contract specifies the equipment to be installed, but you also want to know what sort of specification performance is guaranteed. Remember this fine point — you're buying a performance package, not a bunch of hardware.

- ✔ **Know who's liable.** Figure out who's providing the warranty — the contractor or the manufacturer. Ask whether the contractor is warranting the installation. Figure out who's going to pay for collateral damage and what happens if a faulty system wreaks havoc inside your house. Know who's liable if a contractor falls off your roof. Check with your homeowners' insurance to see whether you're covered for problems such as freezing damage. Will your insurance go up? What are the deductibles for mishaps?

- ✔ **Look out for red tape.** Find out whether you need a county building permit or inspections.

✔ **Decide how you want to establish that the contract is completed properly.** If you pay cash after the hardware is installed, what assurance do you have that your system is going to operate according to the contract specification? What if it doesn't? What if your energy savings are nowhere near what they should be? This can become a nasty blame game, especially if you haven't thought it out beforehand. Make sure that you and the contractor have a specific way to measure system performance and then get that method into the contract.

✔ **If you're going to sell your house, find out what sort of documentation you need to pass along to the new buyers.** Are they expecting a home warranty? Will a home-warranty company cover your solar addition? They'll surely want the instruction manuals.

# Heating Things Up in Your Old Hot Tub

The old style hot tubs are notoriously inefficient because they lack insulation. The biggest energy consumption with a hot tub occurs when the pump is running and the cover is off. When you're using your bubble generator, the water is cooling very quickly as well. Hot tubs can use over 10 kWh when they're cranking away at full speed. That's mucho grande, maybe the champion in your entire household.

If you have a hot tub, the best bet for keeping it that way is to use the system shown in Figure 12-1.

You need a collector that can achieve over 105°F water temperatures — this implies a flat panel or some other insulated collector. A temperature probe is located in the collector fluid, another in the hot tub itself. The delta controller (*delta*, in engineering parlance, means *difference*) decides when to open the valves and run the pump.

Note that a big difference between this system and most others is that you're not interested in storing the most amount of heat for later use; you only need to keep the hot-tub water regulated at a specific temperature. Too hot is just as bad as too cold. Therefore, this system is inherently inefficient because you'll rarely be using it to its utmost capacity. In a way, this is kind of a shame.

These systems are expensive, on the order of $1,500 for equipment and another $500 for installation. Because you can buy a nice new, very efficient hot tub for around $4,000, you may want to consider doing that instead.

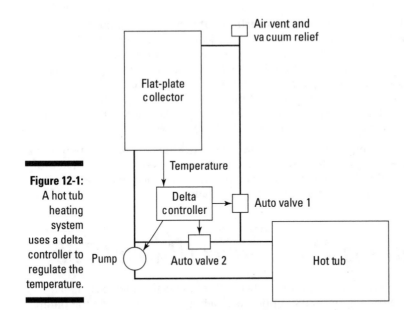

# Installing an ICS Batch System for Mild Climates

If you live in a mild climate, you can install a simple ICS system because you don't have to worry about freezing conditions and super hot conditions. Even if these conditions are rare, you can still install an ICS, if you take a few precautions.

Figure 12-2 shows a simple and effective system for preheating the water that goes to your existing domestic water heater. Over the course of a sunny day, the water in the collector heats up from solar radiation. Due to the insulation, heat doesn't exit the system anywhere nearly as quickly as it enters. Because of the thermosiphon effect, hotter water migrates to the upper copper tube (even when no water flows, the heat will flow) so that the water that ultimately flows down into your water heater tank (when a faucet is opened in the house) is the hottest water from the collector.

Whenever somebody opens a hot water tap in your house, that much water is pumped through the collector, into your domestic hot water tank.

Collectors heat the water directly, so they're a significant part of ICS. Figure 12-3 shows two different versions of typical solar collectors. The collector on the right-hand side shows the more common arrangement. It offers some distinct advantages. The large diameter, black finished, copper tubes beneath the glazed cover (and insulated from the frame) are connected in series so that water flows from the bottom to the top (such collectors are mounted at an angle). Each of the copper tubes can typically hold 10 gallons of water.

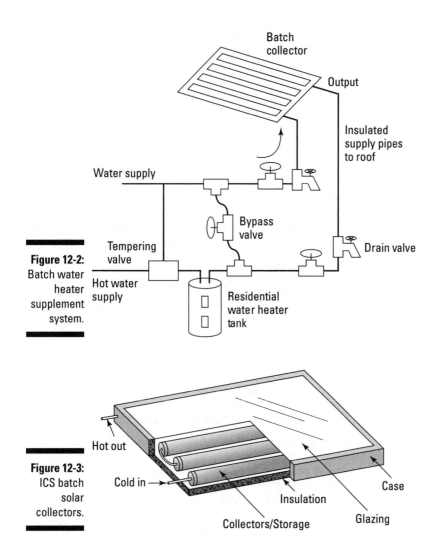

**Figure 12-2:**
Batch water
heater
supplement
system.

**Figure 12-3:**
ICS batch
solar
collectors.

A 3' x 8'–unit holds 30 gallons of water and collects about 22,000 Btus for an average North American day. Cost is around $1,500 for the collector, about $2,200 if you add in the cost for pipes, installation hardware, and labor for installation. Larger units are also available: A 4' x 8'–unit holds 50 gallons of water and can collect 30,000 Btus per day. Cost is around $1,700, plus installation. Or you can expand to two batch collectors by plumbing them in series, but you need to mount the second system physically above the first in order to ensure the proper thermosiphon effect.

Complete kits include all the valves, plus the collector and its associated mounting hardware. If you do opt to do it yourself, give some serious consideration to how you'll lift the collector (even empty, they weigh a lot) wherever you plan on mounting it. The collector can weigh upward of 500 pounds when full of water (which means the units can withstand winds of up to 180 mph!). Make sure that your roof can take this weight. You don't want to find out the hard way that it can't.

If you mount heavy collectors close to the edge of your roof, the rafter load will be easier to handle because the load will be directly over a load-bearing wall. And if you mount one of these collectors right near your water heater, you can use as little as 8 feet of tubing to complete the system.

Here are some considerations to keep in mind concerning the valves:

- **Draining the system:** Valves drain the water when freezing is a possibility for more than a day or so. Note the bypass valves, which allow water to bypass the solar collector and revert to normal, hot water tank operation. The valves may be controlled either automatically or manually.

   Heat is a problem as well. These collectors can burst if the water gets too hot, so drain the system when experiencing extremely hot, sunny conditions. Or turn a hot water faucet on in the house (just a little will do) to be on the safe side. If you don't have a temperature probe in the water line up at the collector, you aren't able to tell how hot the water is. When in doubt, drain the system. Consult the user's manual for your collector to get some advice on when this is appropriate.

- **Locating the drain valves:** Make sure to locate the drain valves where children won't open them up and get scalded. For safe drainage, locate the drain valves outside; they may look exactly like a hose faucet. You may want to consider some kind of locking valve that requires a key to open.

- **Controlling water temperature with a tempering valve:** The tempering valve is critical. It mixes cold water with the heated water from the collector when the collector water exceeds a certain temperature. This prevents scalding water from entering your household plumbing system. Always use a tempering valve in your system and never buy a cheap one.

- **Testing water temperature:** You can always find out how hot the water in the collector is by sampling from the downstream drain valve. Learn your system performance, and you'll be rewarded.

# Installing a Drainback System for Most Climates

Indirect, active, closed-loop drainback systems, shown in Figure 12-4, are an excellent choice for most climates — except those that receive a lot of snow and get really cold. They're the best choice in hot climates, though they're more expensive to install than ICS. The advantage is that you have no freezing danger. (See the earlier sections on system types.)

The controller reads two temperature sensors and then determines when to pump fluid through the collector. When the pump is off, the fluid in the collector and feeder pipes drains back into the drainback tank. (This requires that the collector be mounted higher than the drainback tank, which may be a problem in some applications.) Fluids other than water work in the closed-loop heat exchanger circuit; however, water's best because it doesn't degrade when it gets hot (unlike anti-freeze solutions).

Most manufacturers offer complete kits for these systems, so you don't need to calculate the system parameters. These kits work very well; I highly recommend you stick with a kit.

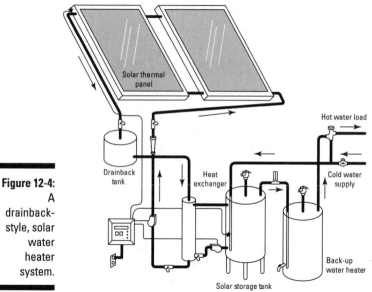

**Figure 12-4:**
A
drainback-
style, solar
water
heater
system.

Here are some installation considerations to keep in mind:

✔ **Raising the liquid up to the collectors sometimes takes a lot of force, so your pump must have sufficient head pressure.** (Measure head pressure by the difference in the vertical height of your collector, and the location of the heat exchanger tank.) You can use a PV module to power the pump, but this is sometimes inadequate because there may not be enough sunlight to yield enough power. On the other hand, if there's not enough sunlight to run the pump, there's probably not enough sunlight to provide meaningful heat. PV power generally works well because you can forego the need for a controller; when there's enough sun for heat, there's enough sun for electrical pumping.

✔ **Installation must ensure the closed-loop system drains quickly and completely.** Locate drainback tanks as high as possible but low enough to ensure the entire weather-exposed parts will drain back. The less distance between the drainback tank and the collector, the less head pressure the filter needs, which means less power consumption.

✓ **Pipes should never be horizontal, so ensure a minimum of 15° slope.** Because purified water is best for the closed-loop system, these systems are prone to freezing. When mounting the pipes, making sure they all have a slope so that they'll drain by force of gravity is critical. Sometimes pipes sag, and the low points retain some water, which can cause the pipes to burst in a freeze. The collector must be mounted so that it, too, will drain completely.

✓ **Don't use 90° bends.** Instead, use two 45° bends staggered apart.

✓ **Larger diameter copper pipe works best.** Use a minimum of ¾".

# Installing an All-Weather, Closed-Loop Antifreeze System

Closed-loop, active, antifreeze systems are the most versatile and failsafe of all the systems. Most new solar homes feature a variation of one kind or another (see Figure 12-5).

A special water-heater tank incorporating a heat exchanger works in conjunction with one or more flat-plate, roof-mounted collectors (see Chapter 10 for details on flat-plate collectors). Glycol, or some equivalent antifreeze fluid, fills the collector and associated routing pipes. A controller measures the temperature in the collector fluid, as well as the temperature in the hot water tank. When heat's available for transfer, pumps are activated to move the collector fluid. The hot water heater tank also has a backup means for independent heating, either gas or electric.

The flat-plate collectors can be mounted in almost any configuration, at great distances from the exchanger. The closed loop is always full of fluid, so the pump pressure requirements are much easier than those for a drainback system (you don't need to worry about head pressures).

The pump can be very small, with very little head pressure, which means you can opt for lower power and better efficiency. Running these types of pumps from a PV panel is practical, which increases efficiency even more. (At night, when no sunlight can power the PV panel, there's no hot water to be pumped anyway, so it's a good match.)

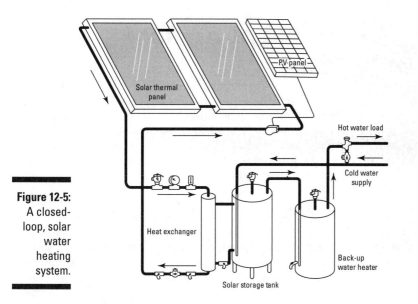

**Figure 12-5:**
A closed-loop, solar water heating system.

You can get complete kits. Installation of the parts is no more difficult than for other systems, but charging the closed-loop isn't straightforward, and doing it incorrectly can damage the system and give you inefficient performance. If you're going to DIY, you need to understand how to fill the closed loop with fluid (there can be no air pockets, or performance suffers).

The major limitation of these systems is the tendency of the antifreeze fluid to degrade over time. When the fluid gets really hot (it's always in the collector, at all times) it degrades quickly, resulting in inefficiency and buildup of deposits on the interior walls of the pipes and collector. For a properly maintained system, you should only need to change the antifreeze fluid every ten years. Preferably, have a qualified serviceperson inject new fluid into the system — it has to be done just right.

The best way to prevent the fluid from overheating is to make sure that the fluid in the closed-loop is circulating at all times when it's sunny out. Therefore, you can get really hot water in your domestic tank, and a tempering valve is critical. You need to use copper pipe — it's the only material that can withstand very high temperatures.

Make sure that the copper pipe is well insulated wherever children may touch it. The pipe can get extremely hot.

# Taking a Quick Peek at Radiant Heat Floor Systems

Installing a radiant heat floor system (see Figure 12-6) is almost surely not a do-it-yourself project, but it merits elaboration because it's such a complete and effective way to use solar energy water heating. A snaking closed loop of metal or plastic tubing runs beneath your floor. When hot water flows through the tubing, the heat radiates upward through the floor and into the room. (For more detail on radiant heat systems, see my book *Energy Efficient Homes For Dummies,* Wiley).

You can use solar heated water to literally heat your home off-grid. Radiant heat, without solar, costs much less than forced-air heating for efficiency reasons. But when you supplement your radiant heat system with a solar water heater, you can drive your heating costs down to nearly nothing. Of course, sunshine isn't very reliable, particularly on the coldest nights when you most need heat, so you can't use solar heating exclusively; it can only be a supplement. But it can be very effective, plus it's also the most comfortable way to heat a home.

You can use any type of water heating system with a radiant floor system, but the capacity of hot water that you use goes up dramatically with a radiant heat floor system. So the attraction of a solar system also goes up dramatically, because you get much cheaper hot water per Btu.

At the very least, radiant heating reduces your carbon footprint. When you combine it with solar, the pollution savings can be impressive.

The engineering is complex, and the installation is clearly not for the faint of heart. There are technical problems, of course, but the systems have been in use for a long, long time. New technologies are making these the system of choice for a lot of homes.

Here's why: With conventional forced-air systems, hot air comes in through the vents and immediately rises to the ceiling. That's not where you want it, so you need to either pump in more heat than you really need (inefficiency) or use overhead fans to move the air back down (inefficiency). Moving air makes you feel colder, and you get stuck listening to blower noise as a big machine goes on and off all night. Furthermore, heated air dries out very fast, so your lips dry up, and your skin gets tight.

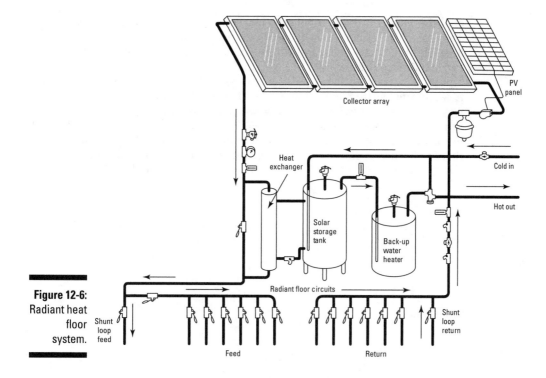

PV
panel

Collector array

Heat
exchanger

Cold in

Hot out

Solar
storage
tank

Back-up
water
heater

Radiant floor circuits

**Figure 12-6:**
Radiant heat
floor
system.

Shunt
loop
feed

Shunt
loop
return

Feed

Return

With radiant floors, the heat starts at ground level and rises naturally, which is much more efficient. There's no blower noise, wind chill is nonexistent, and you don't have to mess with HVAC filters. The big benefit is that the heat is in the room — the floor and furniture — not just the air. You can adjust your thermostat to a lower temperature in a radiant house and achieve the same comfort level because the floor and furniture are where the heat is. Where you set the thermostat is a question of comfort, not numerical temperature.

The floor of your house gets nice and warm. You can use hardwood floors or tile and slate, and the floor never feels cold. You can go barefoot in the middle of winter, and you don't end up hopping around looking for your slippers.

If you're planning on a room addition to your house, consider using a radiant floor in that room. Your existing HVAC system likely won't have enough capacity to heat an additional room. Adding a solar water heater to your house and using your domestic heater to heat the radiant floor in the addition works wonders, and it's usually cheaper than adding another small HVAC system.

You can also cool your house with radiant flooring. It doesn't work quite as well as heating, but if you have solar panels, you can use these at night to cool the water that's already in the closed loop of the radiant floor system. The reason the collectors will cool is simply because they have so much area, and the heat will escape into the cool, nighttime air. This is especially true if a breeze is blowing.

A geothermal system in conjunction with radiant heat is also an excellent way to go. *Geothermal* is a system of wells drilled into the ground; water is run through the wells, and the inherent warmth of the earth's interior is used in a heat exchanger. In general, geothermal systems work best in climates that are both very hot in the summer and cold in the winter. See my book *Energy Efficient Homes For Dummies* (Wiley) for more details on geothermal.

# Chapter 13

# Breezeway: Directing Wind within Your House

*U*nless you live in a perfect climate, heating and air conditioning make up the largest component of your power bill. Ultimately, what you're buying is nothing more than comfort, because you don't need air conditioning — and most of the time, you don't need heat. But comfort comes from more than just air temperature. Humidity is a factor, and moving air feels much different than static air at the same temperature. By moving air appropriately through your house, you can achieve a higher level of comfort — and save money by using your heating, ventilation, and air conditioning (HVAC) system less, particularly in the summer.

*Note:* To gain the most advantage from moving air, you need to first manage solar heat entry into your house. Chapter 9 addresses sunscreens, window blinds, awnings, and other shading methods. Chapter 8 shows methods for planting trees and bushes around your house to optimize breezes and shading.

## Capitalizing on Natural Ventilation

Each house has a natural ventilation scheme that results from weather patterns, physics, and the layout of the house. You need to understand the natural ventilation scheme in your own house before you decide how to optimize air movement by using active fans or passive effects. *Active* methods include ceiling fans, whole house fans, attic fans, and so on. (I get into fan types in the upcoming "Choosing and Using Your Number-One Fans" section.) *Passive* methods take advantage of heat rising and cooler air settling; because no fans are needed in passive systems, there's no draw on your utility power.

Writing a general prescription for moving air is impossible because house layouts are so different. But in this section, you find out about passive methods (with a little active assistance) that you can tailor to your own floor plan. By exploiting both the chimney effect and prevailing winds, you can passively cool your house well enough to preclude air conditioning on all but the hottest days. The following sections explain how these phenomena work, how you can exploit them to get the best cooling effect, and how to adjust your strategy as the seasons change.

The most efficient ways to move air always complement the natural ventilation scheme and never work against it. This is the key to making passive systems work; cooperation, not competition.

## Catering to the prevailing winds

*Prevailing winds* are the most common wind speed and direction in your area. Typically, prevailing winds come from the southwest in the summer and the northwest in the winter. Some areas have very consistent prevailing winds, but other areas experience changes almost daily. You may already have a good idea what the prevailing wind is at your house, but you can refine your understanding by keeping a log — say once a day for a month (and then one month out of each season, because prevailing winds do vary over the year). In particular, how does the prevailing wind at your house change over the course of a day? Over the seasons?

Prevailing winds dictate the arrangement of fans and window openings that work best in your house. You want to create a kind of mini wind tunnel so that the air follows its natural path, coming in through one window and blowing out the one across from it without getting sidetracked. Here's a typical situation (refer to Figure 13-1):

- If all four windows are closed, you get no breeze in the house.
- If only one of the windows is opened, very little air movement occurs in the house.
- If windows 2 and 4 are opened while 1 and 3 are closed, very little air movement occurs because the two open windows are at around the same air pressure.
- If windows 1 and 3 are both open, a good breeze moves through the house.
- If all four windows are open, you get good air movement.

With windows 1 and 3 open, aiming a fan out window 3 enhances the natural scheme. However, if you mount a window fan aiming into the house at window 3, it works directly against the natural ventilation scheme. You may end up with no air movement at all (see Figure 13-2), and you draw utility power, which is about as inefficient as you can get.

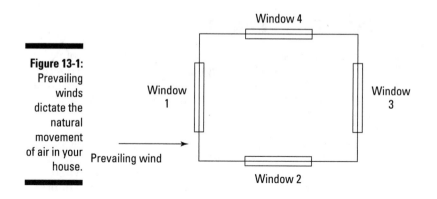

**Figure 13-1:**
Prevailing
winds
dictate the
natural
movement
of air in your
house.

Window 4

Window
1

Window
3

Prevailing wind

Window 2

The prevailing winds blow toward window 1, so a fan at window 1 should blow into the room. If you set a fan in the middle of the room with windows 1 and 3 open, with the fan facing window 3, you enhance the natural scheme, but not nearly as efficiently as when the fan is right in the window.

Consider the different situations outside of the windows. If a particular window has a nice big tree overhead with lawn directly outside, air being drawn into the house through that window is cooler than that from a window situated over concrete in the direct sunshine. In Figure 13-2, a big tree over window 1 works the best.

## Letting the heat rise: Chimney effect

*Chimney effect* describes how hot air rises. In a closed room, the temperature at the ceiling is always higher than that on the floor because hot air is less dense than cold air (hot weighs less than cold). The difference in air temperature can be more than 15°F. You can achieve cooling, without any prevailing breezes at all, by arranging vents in your house so the heat can escape in a fairly direct path. Note that the openings are at different heights, which is key. Figure 13-3 shows a possible arrangement.

Always work with the natural ventilation scheme, never against. If you install a whole house fan in the attic vent, as shown in the figure, you need to direct the air up, into the attic, never back down into the house. (For more info, see the later sections titled "Whole house fans" and "Installing a Solar Attic Vent Fan.")

Many different types of vents allow you to take advantage of the chimney effect. Add some of the vents in the following list if you see the need (see Figure 13-4): ridge vent, roof vent, gable vent, soffit vent, whole house vent (with fan), and kitchen and bathroom exhaust vents.

## Optimizing air movement

To lay out your air-movement strategy, first draw a rough floor plan of your house, marking all the windows, doors, skylights, vents, fans, and so on. Draw your attic, with all the vents and openings. Then, using what you know about prevailing winds, try to figure out what the natural ventilation scheme is for both inside the house and the attic.

On paper, experiment with different combinations of doors and windows and vents to determine how best to achieve comfort without using fans and other active devices. Then imagine using fans to forcibly move air. Where would one single fan work the best in conjunction with windows and doors? How big does that one fan need to be? One small fan in the right location can easily produce more comfort than a huge fan in the wrong spot.

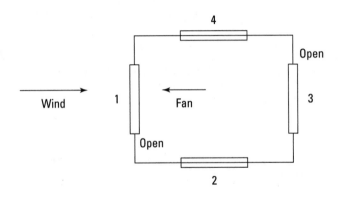

This destroys the natural ventilation scheme

**Figure 13-2:**
Locating fans to enhance the natural ventilation scheme.

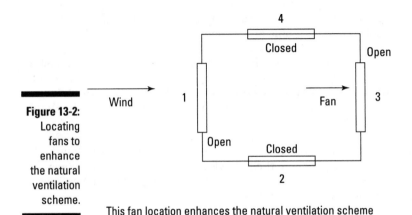

This fan location enhances the natural ventilation scheme

Make sure that you follow some general ventilation rules:

✔ To increase the speed of the breezes, use smaller openings for the inlets and larger openings for the outlets.

✔ Make the air move over as long a path as possible. Windows a few feet apart don't do much. Opening all the windows at the same time doesn't do much, either.

✔ Hot air rises, so locate an attic vent at the highest part of your attic.

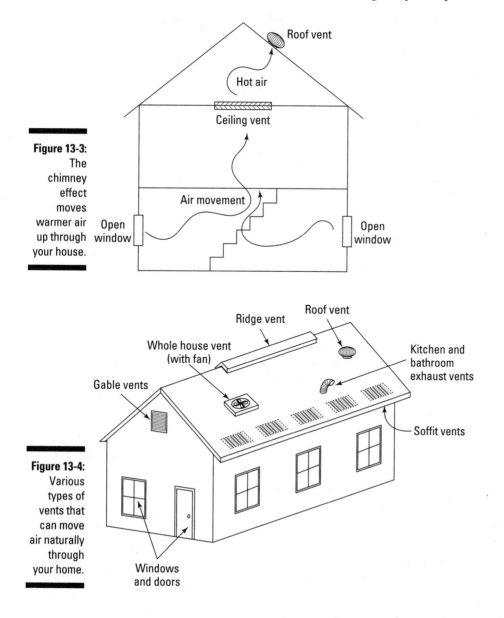

**Figure 13-3:**
The chimney effect moves warmer air up through your house.

**Figure 13-4:**
Various types of vents that can move air naturally through your home.

✔ Try to draw in air from cooler outside areas, but note that doing so may not be practical due to prevailing winds.

✔ Don't open or close all windows at the same time; optimizing your breezes takes a strategy that ultimately boils down to trial and error.

✔ Determine a daily routine. What works best in the morning is rarely what works best in the afternoon or at night.

## Dealing with seasonal variations

Requirements in winter are usually diametrically opposed to those in the summer, which makes the strategizing trickier. In the winter, you want to retain heat as much as possible, whereas in the summer, you want to get rid of heat. Most houses are designed to optimize one season only, which creates problems during the other seasons. You need some means of opening or closing vents to optimize for both winter and summer.

Consider the following scenarios as you play around with your house layout drawing in search of the best solutions.

✔ **Summer:** Dehumidifiers work well with fans, window blinds, and air-movement techniques. Without the air conditioning on, you can open and close windows and vents at will. You want maximum, cool breeze through the house, hopefully from a window that's under a big tree so the air's cooler. You also want to maximize the breeze in your attic, which you can accomplish by locating a vent at the highest point in the attic (to capitalize on the chimney effect). When the AC is on, close all windows (and all window coverings as well, to increase insulation and prevent solar heat from entering the house).

Consider how heat moves in your house when the air conditioner's on. If you can arrange a fan to blow air over your body while the AC is on, you can turn the thermostat up and get the same comfort effect; well-directed small fans are always a good idea in an air-conditioned house.

✔ **Winter:** Close all windows so that prevailing wind isn't a factor inside the house; however, air flow still matters in the attic. You want to close off the attic space in order to retain heat up there. You also want to direct the heat to where you spend most of your time, probably the kitchen and family room. Humidifiers can increase comfort and allow you to turn down the thermostat a few degrees, letting you come out ahead in terms of power draw.

If you have a roof vent that purges hot air in the summer, cover it up in the winter. Heavy plastic garbage bags and duct tape work well (but aesthetically minded neighbors may start a petition, so use your common sense).

✔ **Fall and spring:** Depending on the weather, use techniques from either summer or winter. Generally, you don't have to do much of anything in spring and fall because of the comfortable temperatures.

# Choosing and Using Your Fans

After you determine the natural ventilation scheme in your home (see the preceding sections), you can enhance the effect with the use of active fans. Like anything else, some fans are better than others. What you want is to achieve the highest efficiency at the lowest cost (both operating and equipment costs). In this section, I present some general rules for fans.

## Viewing fan efficiency and operation

The *efficiency* of a fan is the amount of air it moves divided by the power it consumes. Fans are rated for how much air they can move in a given amount of time. The most common spec is cubic feet per minute (cfm). You can calculate how many cubic feet are in your house, room, or attic (length × width × height) to find the proper size fan. Be aware, however, that the fan is measured in ideal conditions of air flow, which you probably won't be experiencing. Buy a fan about 20 percent larger than your calculations suggest. But also be aware that if you size the fan wrong, it's not really a big deal. Just doing something is better than nothing.

Long, meandering hallways impede air flow, so make your air flow routes as straight and clear as possible. Consider the entire airflow path and make the flow as easy as possible. You lose efficiency if you

✔ Aim a fan directly at a wall

✔ Set a fan near a window that's only cracked open

✔ Aim a fan through a dust-coated grate

The most efficient movement of air occurs when a fan has a *cowling.* (Cowlings prevent air around the tip of the blades from circling about the outside of the blades — they're used extensively for jet engines.) A fan without a cowling allows air to loop around the outer edges (see Figure 13-5); a cowling prevents the looping and forces the air forward, moving much more air for the same power draw. An enclosed window fan with a cowling is the most efficient setup of all. (If all you're after is some air movement in a closed room, you can skip the cowling. But if you're interested in moving a large quantity of air through your house, it becomes more important. Most heavy-duty work fans come with cowlings.)

*Thermostats* are switches that open or close at a set temperature. They come in handy when you set up a fan in a remote location, such as an attic. When the attic reaches a certain temperature, the thermostat (and the fan) activates. When it's hot in the summer, you get ventilation. In the winter, the fan never comes on. You can get a *thermostat* in most hardware stores for around $15. You have to do some AC wiring, which means there may be a shock hazard if you're not careful. You may want to hire an electrical contractor for the job.

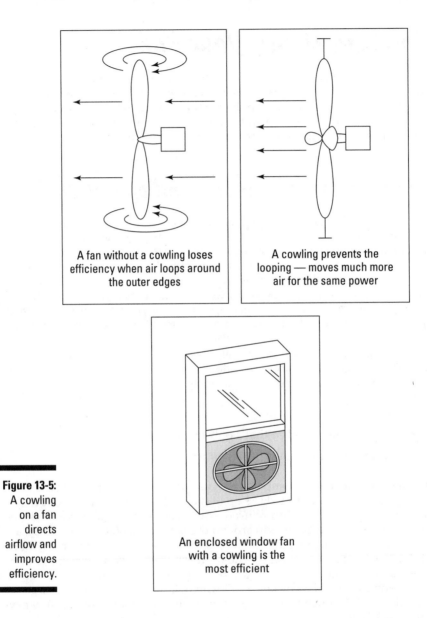

A fan without a cowling loses efficiency when air loops around the outer edges

A cowling prevents the looping — moves much more air for the same power

**Figure 13-5:**
A cowling on a fan directs airflow and improves efficiency.

An enclosed window fan with a cowling is the most efficient

Never leave a fan on in an enclosed room when no humans are present. All fans are heaters. They dissipate power, giving it off as heat. If you put a fan in a closed room, the air temperature will rise — it's a fact of physics. You may feel cooler because of the breeze, but the temperature rises. Leaving an oscillating fan on all day long in an enclosed house while you're at work only wastes electricity and heats your house.

# Considering fan types

Many fans are available at a wide range of prices. Choosing the right unit for your needs can ensure you get the proper effect you're looking for.

### Oscillating fans

Oscillating fans move back and forth. These fans work only for convective cooling, and they're dubious at that. They do a good job of evening out temperature variations in a room, but that's rarely what you're after because you spend a majority of time in any given room in only a few locations. For instance, in your family room you spend most of your time in your favorite recliner.

If you want to achieve the most convective cooling for the least cost, get a small, stationary fan and aim it directly at your uncovered body (I'm not suggesting complete nudity, although complete nudity does have its merits). Position the fan close so that you get the most benefit from the air movement. A big oscillating fan on the other side of the room is inefficient and may even make things worse by stirring up hot air near the ceiling. It's best just to leave that heat up there where it's minding its own business and not bothering anybody.

### Box fans

Box fans are portable units you can move around as the need occurs. Most have some kind of cowling, some better than others. Blade design also determines the efficiency of the air flow. You can find fans that are impressively efficient, but keep in mind that paying extra for a good fan may not be as worthwhile as simply using a smaller, cheaper fan the right way in the first place.

If you have enough space, a large, lower-powered fan is better than a small brutish one. The larger the blades (not power, blade size), the more slowly the blades can move to achieve the same amount of air. Fan noise is important. Cheap ones are loud, clanky, and inefficient. Good ones make a smooth, even sound.

### Window fans

Window fans come with a sheet metal mounting arrangement that fits right into an open window and seals around the edges (a perfect cowling). Most of these fans are very efficient. The best way to move air in your house is by using a good window fan mounted into the most appropriate window, probably upstairs on the downwind side of your house (the fan aims out the window). One small window fan mounted properly can do the same work as a number of large fans scattered about the house. (See the earlier "Catering to the prevailing winds" section.)

### Exhaust fans

Consider exhaust fans in bathrooms and kitchens. When you turn on a bathroom fan without any bathroom windows open, the exhausted air is drawn from the house, and outside air is drawn into the house through whatever openings are available (leaks in the insulation, open windows, and the like). Exhaust fans shouldn't be used when the HVAC system is running, but if you need to, open a small window nearby so that you can control the air movement.

If you're cooking in the summer when the AC is on, exhaust fans can make a big impact on the comfort in your house. This is true especially if you're boiling water, because not only does that heat the air, but it also humidifies. Run the fan on high, but crack open a nearby window so that you don't pull expensive, cooled air out of your house. Also, when you're finished cooking, run cold water from your tap over the heated pots and pans to wash the heat down the drain instead of letting it slowly seep into the house.

### Ceiling fans

Ceiling fans may or may not bring outside air into the house, depending on how you've set up vents. They can accomplish two things: convective cooling and reversing or enhancing the chimney effect. Here's how to use your ceiling fan to get the most benefit (see Figure 13-6):

- ✓ **Create some convective cooling:** All you need for convective cooling is air movement, and it doesn't really matter how you achieve it. With a ceiling fan, you can get convective cooling by running the fan in either direction. If you run the fan so that air is pushing directly down at you, it will feel entirely different than if the air is being pushed up, around, and then down against the walls. The latter is a more indirect movement and may feel more comfortable by virtue of being less imperative.

- ✓ **Draw the heat up:** If you have a vent (or an open window near a ceiling), you can enhance the chimney effect by running a ceiling fan in the reverse direction (aiming the air flow upward, in other words).

- ✓ **Reverse the chimney effect by pushing hot air down:** In the winter, you can use a ceiling fan to push hot air back down into the room. In a small room, it won't make much difference which direction the fan is running, because the air will simply circle around; the net effect is to equalize the temperature in the entire room. In a large room, push the air downward. Note that in this case, if you're sitting directly below the fan, convective cooling from the air movement may exceed the warming effect. Experiment and you'll find an optimum.

In the summer, whether a ceiling fan is really beneficial is an open question. You want hot air to rise to the ceiling and stay there; however, when you run a ceiling fan in either direction, you're stirring up the hot air and moving it around the room. On the other hand, you get a very nice convective cooling effect from a ceiling fan, which may more than offset the warming effect.

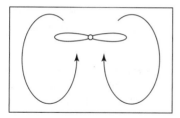

Closed room:
Enhances chimney effect

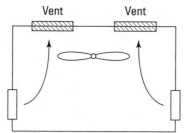

Vent        Vent

Venting to achieve
maximum cooling

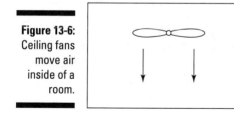

**Figure 13-6:**
Ceiling fans
move air
inside of a
room.

Winter: Pulls heat from
ceiling, into room

A well-positioned ceiling fan can allow you to lower your thermostat by 4°F and achieve the same comfort level. Ceiling fans work best when the fan blades are 7' to 9' above the floor and 10" to 12" below the ceiling. Cramped ceiling fans are inefficient.

You may need several fans in a large room. Aesthetics will likely dictate the choice of size, but the bigger blades are quieter because they move more slowly to achieve the same airflow.

The maxim *you get what you pay for* is especially true for ceiling fans. Cheap ones make noise, are inefficient, and don't last long because they have shoddy bearings. In other words, cheap ceiling fans are very expensive, particularly if you factor in the time and effort it takes to install one.

### Whole house fans

*Whole house fans* are mounted in the ceiling between a high point in your house and the attic. They draw air through open windows, exhaust it up through the attic space, and cool the attic space. (Any air movement through the attic space will cool it because the air you're drawing in is almost always cooler than the air in the attic space.)

You never run a whole house fan with the HVAC on and run one rarely in the wintertime. But in the summer, when the outside air temperature cools down at night, a whole house fan works better than any other air-movement scheme.

A great advantage is that you can open any window in the house and get air movement in that room as long as the door is also open. You can open some windows more than others and regulate the relative effects.

In the summer, when the outside air is cooler than the inside air, you turn on the fan when the sun goes down and leave it on all night. (If you've been running your air-conditioner all day and the inside of your home is cool, running a whole house just purges the cool air and draws in hotter air.) In the morning when the sun comes up, shut the fan off and close all the windows and drapes in the house to keep the most heat out. In the winter, it's a good idea to cover a whole house fan with some kind of insulation so that you don't lose a lot of heat up into the attic.

The easiest way to cover your fan for the off-season is to unplug your whole house fan and simply cover the whole unit (on top, from the attic) with a thick, old blanket from a thrift store.

Some house fans come with metal vents that automatically open or close when the fan is on or off. The insulation properties of these are very poor, and they generally leak a lot as well.

Here are some mounting tips:

- ✔ Whole house fans make a lot of noise because they're big and powerful, and because they need to be mounted in a central location, the noise is prevalent in the entire house. Mount them on rubber gaskets to reduce the noise.

- ✔ If your attic isn't well vented, a ceiling fan will move very little air for the amount of power it consumes. Go up into your attic and check to make sure that you have plenty of vents. If not, either make some vents or don't use a whole house fan. And make sure the vents aren't coated with dust, which impedes flow.

### Attic vent fans

As opposed to a whole house fan, an attic vent fan draws air in from one part of the attic, and out another. There is no air movement in the house below, so attic vent fans can be run when the house is closed up (for example, when the air conditioner is on). See the upcoming "Installing a Solar Attic Vent Fan" section for project advice.

# Directly Powering Tabletop and Ceiling Fans

You can use PV modules to directly power tabletop fans (portable units) and ceiling fans. Tabletop fans that run off of 12VDC are available (look for them on RV, boat, and camper Web sites). This method is also practical for gazebos, porches, outbuildings, and other applications without AC power.

The advantage of solar power is that you don't need to pay for expensive electrician labor or county building permits to install hardwiring. For example, in a remote gazebo, you'd have to first get a county building permit to run household electrical power to the site. Then you'd have to dig a deep trench for the electrical conduit and hire an electrician to run the conduit and connect to your fuse box. The electrician would also have to configure switch boxes, switches, and other code wiring, finally culminating with a junction box where the fan needs to hang. The cost of such a project can run well over $2,000, not including the fan itself.

On the other hand, a complete do-it-yourself solar fan system might cost $300 (see the next section).

Greenhouses are an excellent application as well because they generally don't need electrical power for any reason other than driving fans to cool the air in the summer. Solar-powered fans are ideal because there's no need for expensive electrical wiring. Plus, they're safer because they run off of low, DC voltages. Greenhouses are often wet and humid, creating shock hazards with conventional wiring schemes.

# Installing a Solar Attic Vent Fan

Attics can get to be over 160°F in the hot summer sun. This heat migrates down via conduction through your ceiling and into your house. Even in the middle of the night, a poorly ventilated attic stays very hot. If you can somehow manage to continuously purge the air in your attic with outside air, your entire house will be much cooler. This idea is especially true of old houses with poor insulation.

One way to purge an attic is with a whole house fan (see the earlier "Whole house fans" section). If the weather's not too hot out, a whole house fan can keep air moving through the living space and purge heat from the attic above. But you can't run them when the air conditioner is on because they require the house to be opened up, and they don't have the desired effect when the outside temperature is hotter than the inside. However, attic vent fans can be run anytime, regardless of whether the house is opened up or closed off. In particular, when you're running an air conditioner, you want an attic vent to cool the attic as much as possible because that will decrease the work load on the air conditioner (which is very expensive to run).

Cover up your attic vent fan in the winter to prevent natural ventilation and keep heat trapped in the attic space.

## Opting for solar

A solar attic vent fan can be installed anywhere and requires no county permits or electrician labor. Figure 13-7 shows a solar attic vent system.

The fan runs whenever the sun shines on the PV modules; this makes perfect sense because when the sun is the brightest, you want the most air pumping. The unit comes on only when the set temperature is reached. For the cost and complexity, a solar vent fan may be one of your best solar investments in a hot climate.

**Figure 13-7:** A PV module powers the attic vent fan, allowing hot air to escape when the sun shines its brightest.

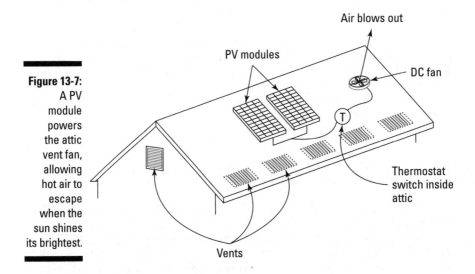

Air blows out

PV modules

DC fan

Thermostat switch inside attic

Vents

## Choosing the best location

Attic vent fans work any number of ways. Figure 13-8 has the fan mounted on the roof and in the gable. (See "Letting the heat rise with the chimney effect" for more on vent types.) Either scheme works essentially the same, the mounting choice depends on ease of installation. The gable scheme is also more weather-proof, because rain isn't apt to fall into the attic space through the opening, and leaks around the seals won't cause problems like they will up in the roof.

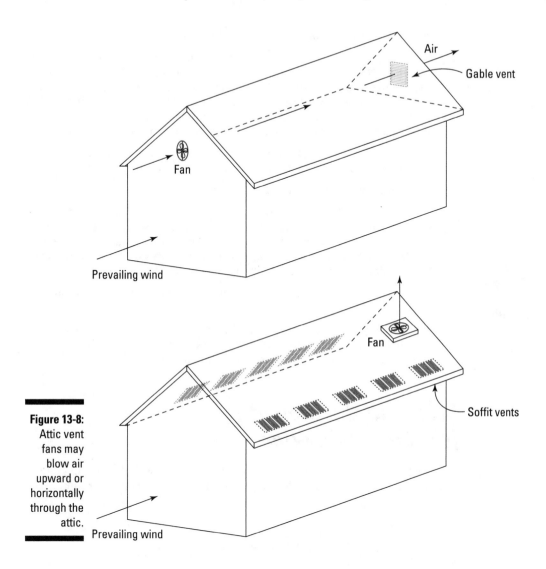

**Figure 13-8:**
Attic vent
fans may
blow air
upward or
horizontally
through the
attic.

For the most efficient setup, place the fan at the apex of the roof, where the most heat collects due to the chimney effect. Note in the figure that the rooftop vent is on the downwind side of the house to complement the natural ventilation scheme from prevailing winds. The following steps explain what to do as you choose your ideal setup:

1. **Determine your attic's natural ventilation scheme.**

    Go up into your attic and find the vents and openings designed for ventilation. Make a rough drawing and an air movement analysis. (See the earlier section titled "Capitalizing on Your Natural Ventilation Scheme.") Identify a location where it makes the most sense to locate an attic vent fan.

2. **Look for ways to improve the natural ventilation scheme.**

   For example, if you only have a few vents and they're in spots that don't enhance prevailing winds or the chimney effect (which says that heat rises), putting in a few more vents may be easy. In particular, most houses could use a vent or two near the top of the roof to enhance the chimney effect. You can get inexpensive vents from most hardware stores, and they're relatively easy and foolproof to install.

   Consider what the vents will look like from the street. You may want to locate roof vents so they won't be visible from the street level.

3. **Determine the best location for the fan.**

   In keeping with the rule to always enhance the natural ventilation scheme, the most logical spot to put a vent fan is usually in a roof vent, near the apex, with a cowling around the fan. Many roofs already have passive vents; these openings are prime candidates for installing a fan, but the dimensions and accessibility may preclude this.

   Always try to locate a fan downwind in order to equalize ventilation throughout the attic.

   Also pay attention to available sunlight at potential locations on your roof. If you can't find a place that gets evening sun, you may have to buy components separately, as I explain in the next section.

## Selecting the type of fan for your application

Attic vent fans come in two forms: one-piece units and distributed units. A one-piece unit is built as in Figure 13-9. Total cost for a one-piece unit is around $300.

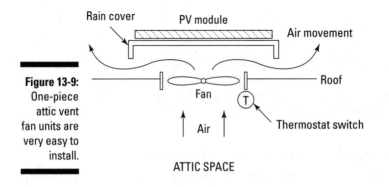

**Figure 13-9:** One-piece attic vent fan units are very easy to install.

One-piece attic vent fans are very simple to install; you don't even need to go into the attic space. Simply cut a round hole in your roof, pull the shingles back, slide the unit up under the shingles, and drop it into the hole. Seal for weatherproofing, and you're done.

Avoid trusses by going into the attic and drilling a hole out through the roof halfway between adjoining joists. Back up on the roof, find the hole, and go from there.

However, one-piece units may present a potential problem: The solar modules are fixed in place on the unit, and the best location for your vent fan may not be the best location for collecting sunlight. (If your solar modules aren't getting good sun, the fan won't be pumping good air quantities.) In this case, you need to go with a distributed-type system.

You can buy separate components for much less cost than a one-piece unit — a separate 12VDC fan powerful enough to do the job costs about $100, and a PV module to run it costs another $130. However, the installation is more expensive and difficult. You need a hood of some kind over the fan to prevent rain from getting into the attic, or soaking the fan itself. You could potentially install the fan on a vertical surface, behind a grate, and avoid a lot of installation problems, like cutting through a roof (which always entails risk or leaks).

Get better performance by tilting the PV modules to the west so you get the most air movement in the afternoon. Tilting the modules to peak at around 2 p.m. can provide the most comfort in the house.

# Building a Solar Space Heater

You can build a very effective solar space heater for around $400 in parts (see Figure 13-10). Over the course of a sunny day, this system can heat a room at zero cost — there are no pumps, fans, or moving parts. The size shown can heat a small office in the dead of winter (as long as the sun is shining), or you can build smaller units to distribute around your house. They work well in outbuildings (barns, work rooms, and so on) with no power available.

The unit operates on both the chimney and greenhouse effects. The solar collector absorbs sunlight and converts it into heat, which becomes trapped in the collector via the greenhouse effect. The air around the absorber heats, expands, and rises, creating a natural convection current via the chimney effect. Vents at the top and bottom of the collector allow air-looping movement between the collector and your house. Cool air is drawn into the bottom, heated in the collector, and then feeds back into the house through the top vents. In the summer, you need to block off the vents to prevent operation. On a cold night, close the vents to prevent the reverse process from occurring.

The best location is a south-facing wall, but east or west can also be useful, depending on what time of day you want the heat.

Here's a parts list for a 160-square-foot collector that you can build for a total of $417:

✔ 68 ft. of 2" x 6" lumber for the verticals and bottom sill ($45)

✔ 22 ft. of 2" x 8" lumber for the bottom sill ($20)

✔ 130 ft. of 1" x 1" lumber for the glazing and supports ($20)

✔ Ten 8' x 26" corrugated polycarbonate panels (used on porch roofs) for the glazing ($180)

✔ 40 ft. of foam closure molding ($12)

✔ 300 ft.$^2$ of black metal screen for the absorber ($70)

✔ Miscellaneous hardware, including sealant, fasteners, and paint ($30)

✔ Vents ($40)

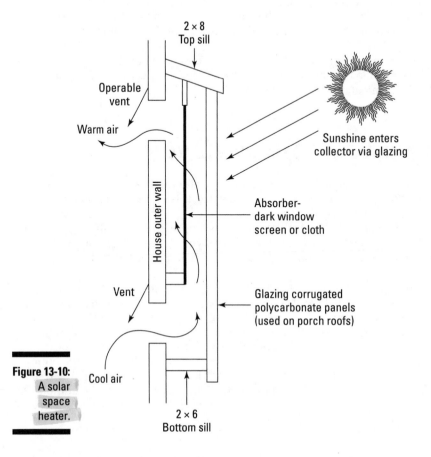

**Figure 13-10:**
A solar space heater.

# Chapter 14

# Solar Pumps, Wind Power, Hydro Power, and Batteries in Applications

*In This Chapter*

▶ Running pumps directly from PV modules

▶ Using batteries when powering small remote cabins, RVs, and boats

▶ Considering wind and water power as alternative solar power resources

*Y*ou can use portable solar PV systems for cabins, boats, RVs, backups, in-home offices, and outbuildings, such as barns, offices, exercise rooms, and so on. The usual advantages apply: reducing both your electric utility bill and pollution. You can also use PV systems to run well pumps and swimming pool pumps. Well pumps are often located in remote locations, so the use of solar PV for power precludes the need to pay for costly power transmission line runs. Solar powering your swimming pool pump makes sense because the pump will run the hardest when the sun is the brightest, which reflects how the pool itself is used. Wind and water power are also solar resources, so I describe in general terms when and how both alternatives should be used and point you in the right direction if you decide to pursue them.

As with any solar application, energy conservation is the first step. Before specifying your system needs, do an energy audit and take the indicated energy conservation measures (see Chapters 2 and 3). A general rule: Each dollar you save on conservation will save you around $5 on the cost of a small solar-power system.

Considering solar alternatives other than electrical PV is also wise. For example, solar water heaters work very well for remote applications. Do-it-yourselfers can build batch-style water heaters for very little cost (see Chapters 10 and 12). Chapter 9 includes a good number of small solar systems that you can use independently of other systems.

# Directly Powering Pumps and Motors

PV panels (see Chapter 4) can directly power DC electrical motors (no batteries required). Chapter 13 details an attic vent fan system. Other applications are similar in nature with regard to system components and operation.

You can purchase individual parts and assemble and install systems yourself, but buying completed kits makes a lot more sense because the pros probably know how to integrate parts into working systems better than you do. The design will be optimum, which is often not an easy engineering task.

## Water supply systems

Solar power is very useful for water supply systems. The most common applications are for household water supplies and agricultural and livestock needs. Of course, using utility-provided water is more economical in most places, but in remote locations, it's simply not available. The pipes and trenches that would be required to get to the desired site are often prohibitively expensive. And in these types of applications, providing utility electrical power — which you'd need to power your own AC pump — is likely also prohibitively expensive, so there's a double whammy.

Solar water pumps can be located anywhere there's available sunshine and a relatively clean water supply (dirty water may be filtered, but filters clog up, and the pump flow becomes constricted, so the application quickly becomes maintenance intensive). The water supply can be a well, creek or river, lake, and so on. (Check on the legal accessibility of your water source before you get too far into the project.)

Water-pumping systems all include a few basic components, as shown in Figure 14-1.

On a sunny day, the PV-powered pump slowly fills the reservoir. In a household application, the storage reservoir is located above the house so that when someone opens a tap, gravity provides the water pressure to the faucets. These types of systems are common for lakeside cabins and homes in the California mountains.

For livestock applications, the reservoir may be nothing more than a ditch or a cattle trough set on the ground. This system also works for crop irrigation where there's no reservoir at all. The water simply feeds directly into the irrigation pipes. Because the pressure varies quite a bit, broadcasting sprinklers are impractical, but drippers work very well.

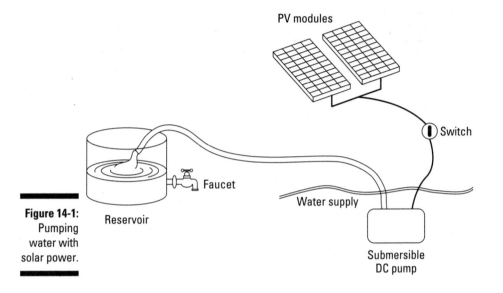

PV modules

Switch

Faucet

Water supply

**Figure 14-1:**
Pumping
water with
solar power.

Reservoir

Submersible
DC pump

Here are the calculations you need to make in order to specify the system size:

- ✔ **Daily water usage, in gallons, both average and maximum:** This amount may vary over the course of the year. For example, some livestock applications don't require water in the winter months at all because the animals are in grocery stores, waiting to take center stage at your holiday banquets.

- ✔ **Available sunshine, in average hours per day, and your need for a consistent water supply:** Can you tolerate a few days of drought? If so, the reservoir may be smaller. If you need water all the time and sunshine is inconsistent, you'll need a larger reservoir, along with larger PV modules and pumps so you can fill the reservoir on sunny days. (See Chapter 5 for info on appraising your solar potential.)

- ✔ **Difference in vertical height between the top level of the water source and the output end of the pump hose:** For agricultural applications, this distance is usually small, on the order of 10 or 20 feet. For some residential applications, this number can be more than 100 feet, which makes the pump much more expensive. This number is called the pump's *head pressure*.

Specifications for pumps list a table of wattages versus pump capacities and heads. For example, see Table 14-1.

| Table 14-1 | Sample Pump Specifications | |
|---|---|---|
| *Head Pressure* | *Wattage* | *Daily Output in Gallons* |
| 3.3 ft. | 80 | 1,500 |
| 9.8 ft. | 80 | 1,300 |
| 23 ft. | 80 | 600 |
| 3.3 ft. | 115 | 2,100 |
| 9.8 ft. | 115 | 1,700 |
| 23 ft. | 115 | 650 |

A pump, plus a matching PV panel, costs around $1,400 for the 80 W version and $1,600 for the 115 W version. You have to add the cost of pipes and tubes, but you don't have to spend much more than that. If needed, a simple float-type switch can turn the pump off when the reservoir is full, but the system will be more economically efficient if you can figure out a way to use all the water that can possibly be pumped.

## Swimming pool pump systems

Swimming pool solar pumps are available, but they're expensive upfront. The advantage is that conventional pool pumps (which are relatively inexpensive) cost a lot to run with utility power, but with a solar system, you'll never have to pay another cent for powering your pool pump. In particular, if you have a solar swimming pool heating system (hot-water collectors) and a large-scale solar PV system on a time-of-use (TOU) rate structure (see Chapter 17), you need to run your pool pump in the afternoon when power rates are at their peak.

A conventional pump system is prohibitively expensive to run in this case. In the long term, you'd be better off with a solar pump system, which would be much more consistent with your green philosophy.

The more sunshine you get, the more the pool pump runs. This works nicely because swimming hours (how much your family uses the pool) usually track the amount of sunshine as well. Table 14-2 shows a typical performance chart for a system. *Note: Back pressure* is a measure of head and piping resistance to flow. If the filter is clogged up and the piping is corroded or blocked, the flow levels decrease as the back pressure increases.

Table 14-2 gives you an idea how much a system costs and which size you need for your particular application.

| Table 14-2 | Swimming Pool Pump Specs | | |
|---|---|---|---|
| *Watts* | *Gallons/Day* | *Pool Size* | *Price* |
| 340 | 17,600 | 25,100 | $5,000 (ouch!) |
| 480 | 23,400 | 33,500 | $5,900 |
| 600 | 27,500 | 39,300 | $7,000 |

# Working with PV Systems with Batteries

You can safely and reliably install off-grid PV systems because the voltages and power capacities are low. But batteries can be tricky (see Chapter 18 for more details on batteries and their operations), so avoid them if at all possible.

## Assessing your power needs

Regardless of the application, the first step is to assess your power needs. You can add up the power consumptions of individual appliances and other loads you need to run, but you also need to define how many loads will be running concurrently and at what times. Here are four specifications of interest:

✓ **Total system capacity, measured in kilowatt-hours per day:** This is the total amount of electricity you need over the course of a day.

✓ **Peak instantaneous power output, measured in watts:** You have a lot of control over this specification because you can run appliances at different times. For example, if your usual routine includes turning on the lights in the morning at the same time you turn on the coffee maker, you can change to a programmable coffee maker that has your coffee ready before you wake up. Or you can run your microwave oven only when all the lights are off, and so on.

✓ **Duty cycle (a measure of how often the system is used weekly):** A weekend cabin used for two days a week has a duty cycle of ²⁄₇, or approximately 28 percent. A system used every day has a duty cycle of 100 percent.

✓ **How many hours of good sunlight a day you can expect:** Estimating with much accuracy is difficult because sunlight depends a great deal on the weather. It also depends on the time of year in which you're interested in using your system. For a weekend cabin in the mountains, used only in the summer, you may expect 8 hours a day of sunshine, and in the thin mountain air, you'll get very good solar exposure. (See Chapter 5 for more details on estimating your sunshine.) For a cabin used twice a month for a few days all year round, the worst case will be in winter, when you may only get four hours a day of cumulative sunshine, even less if it's snowing and the panels are covered up.

Table 14-3 shows a typical sample load analysis chart for a weekend cabin (your table will be different, of course). (See Chapter 2 for a list of typical power draws for different household loads.) In this situation, the duty cycle is around 28 percent (two days per week), and the cabin is used only in the summer months, so the average expected sunlight per day is around eight hours (the weather's rarely cloudy, and the air is thin and cool — excellent for solar applications).

| Table 14-3 | Energy Consumption in a Weekend Cabin | | |
|---|---|---|---|
| *AC Device* | *Watts* | *Hours/Day* | *Watt-Hours/Day* |
| Kitchen light | 60 | 3 | 180 |
| Family room light | 120 | 4 | 480 |
| TV | 70 | 3 | 210 |
| Coffeemaker | 200 | 0.5 | 100 |
| Microwave | 900 | 0.15 | 135 |
| Clock radio | 1 | 24 | 24 |
| Table fan | 15 | 6 | 90 |
| Refrigerator | 20 | 24 | 480 |

Here's how to do some significant calculations from this chart:

- **Total energy needs in one day (in kWh/day):** Total up the Watt-Hours/Day column and divide by 1,000

    1,699 kWh/day ÷ 1,000 = 1.7 kWh/day

- **Daily energy needs adjusted for inefficiency (10%):** Multiply the total daily energy needs by 1.1

    1.699 kWh/day × 1.1 ≈ 1.9 kWh/day

- **Maximum instantaneous load:** Decide which appliances will be on at the same time and add their power draws. (You can control this number by simply restricting the simultaneous use of certain appliances — for example, don't use the microwave and the coffee pot at the same time.)

    1,200 W

- **Duty cycle:** Divide the number of days per week the cabin is used by 7

    2 days ÷ 7 days ≈ 0.28 = 28%

- **Average kWh per day:** Multiply the adjusted daily energy needs by the duty cycle

    1.9 ÷ ⅘ ≈ 0.54 kWh/day

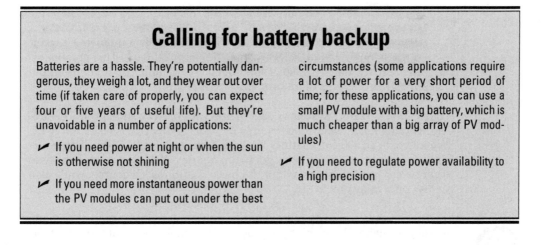

# Calling for battery backup

Batteries are a hassle. They're potentially dangerous, they weigh a lot, and they wear out over time (if taken care of properly, you can expect four or five years of useful life). But they're unavoidable in a number of applications:

 ✔ If you need power at night or when the sun is otherwise not shining

 ✔ If you need more instantaneous power than the PV modules can put out under the best

circumstances (some applications require a lot of power for a very short period of time; for these applications, you can use a small PV module with a big battery, which is much cheaper than a big array of PV modules)

 ✔ If you need to regulate power availability to a high precision

The specifications that define the example system are the following (when you purchase a system, you need these numbers to tell system size and composition):

 ✔ **Solar panel output:** At eight hours of good sunlight per day, the example requires a 67.5-watt solar panel. (Take the average kWh per day of 0.54 and divide by 8 hours.)

   It's best to multiply this number by 1.3, just to be on the conservative side. This yields around 90 watts.

 ✔ **Voltage output:** In the example, it's 12VDC, but you can also use other voltages. (12VDC is the most common because it's what boats, cars and RVs use. The choice is governed by the equipment you use.)

 ✔ **Battery size:** Calculate battery size, which is specified in terms of amp-hours (Ah). Most batteries are 12VDC, but other sizes are also available. For this example, assume you have a 12VDC system. Take the total adjusted kWh/day, multiply this by 1,000 to get kWh/day, and then divide this value by the battery voltage. For the example, this yields 158 Ah:

$$\frac{1.9 \text{ kWh/day} \times 1,000 \text{ W/kW}}{12 \text{ VDC}} \approx 158 \text{ Ah}$$

   In order to get the best performance and economics out of a battery, it's best to over-spec the capacity by 100 percent, which yields around 300 Ah.

 ✔ **Fuse size:** Fuse size is determined by the maximum current draw. Take the maximum instantaneous load and divide by the voltage. For the example, this yields 100 amps:

$$1,200 \text{ W} \div 12\text{VDC} = 100 \text{ amps}$$

> This is a problem. Not only are currents higher than 30 amps dangerous, but the wire sizes also have to be very thick. Note that if you toss the microwave oven out of the picture, the maximum instantaneous power outage is only around 300 watts, which yields 25 amps. This number's practical. It's time to make a solar oven (see Chapter 9) or get a camping stove that operates off of propane.

A basic 12VDC starter systems costs $1,100 with battery and outputs 0.4 kWh per day (which isn't very much energy). The PV panel size is 60" x 24". A 1.6 kWh kit costs $5,000 with battery backup; panel size is 60" x 100". These numbers are for four hours of pretty good sun per day. You may actually be able to get twice these numbers, under the best conditions. Of course, you'll also get less on a rainy day.

You can get more capacity out of your portable solar system (and save money by using a smaller PV panel) by aiming the PV panel directly at the sun at all times. If you can, design a mounting system that makes this easy and practical. The easiest way is to simply set the panel against a wall and move it manually. Or you can mount it on a cart with wheels and turn it occasionally. Aim the panel east in the morning, south during the middle of the day, and west in the afternoon. Or at the very least, change the vertical angle (elevation) a few times per year.

Auto batteries don't work in these applications because they're designed to output high currents for very short periods of time. In household power applications, low currents for long, sustained periods are the norm. Also, auto batteries aren't designed to discharge a large percentage of their capacity. What's needed are deep-cycle batteries, which are the kind used in golf carts.

Many small-scale systems are subject to rebates and subsidies, just like PV systems. See Chapter 20 for details on how to research this subject.

## Powering remote and mobile locations

Standalone PV systems (with batteries) aren't economical compared to using the power grid for your power needs. But where no grid power is available — for example, on boats and RVs or for remote mountain cabins — they're excellent solutions. You can either install a medium-sized system in a fixed format, or you can make your system portable so you can use it in a number of different venues. Or simply move the PV panels and leave everything else fixed. You can get PV modules that are flexible and waterproof. Some PV modules fold up for portability.

Some users are interested in backup power, for when the grid goes down. A cheaper alternative than solar is a gas-powered generator, but these take time to start, make noise, and need ventilation; and they stink, not to mention pollute. For those who are interested in a peaceful, clean environment, solar is the best choice.

It's wise to supplement medium power PV systems with other solar systems, such as hot water heaters and solar ovens (see Chapters 9 and 10). The latter are cheaper per kWh, and by dividing the tasks, you lower your risk that a broken system will disrupt your entire lifestyle.

# Wind and Water Solar Power

Windmills and hydro-power generators are also sustainable solar resources. Both are energy sources derived from the sun, although indirectly compared to solar collectors (PV and hot water). Wind is caused by changes in air pressure brought on by different regions of the globe heating and cooling at different rates due to variations in solar exposure. Rain is caused by evaporation, which results from warming and cooling as well; hydro-power comes from rain storage in reservoirs, creeks, and the like.

The two methods are very similar in the way they generate voltages with spinning turbines. The concept is very simple: When you apply a power source to an electric motor, the shaft spins. On the other hand, if you reverse the process and manually spin the shaft of an electric motor, its two wires output power. The physics goes equally in either direction. Practical designs can be very complex, but at the heart is always the electric motor principle. Inverters are required (just as with solar PV panels) to convert the raw voltages from the rotors into usable power, typically 12VDC because batteries are the norm. Another style of inverter converts into standard 120VAC household current.

Although solar PV panels and water heater collectors work only when there's direct sunshine, both wind and water are available at any time of day or night. Wind, however, can come and go, from minute to minute. Water resources generally don't vary much over the course of a day, but they can vary over seasons and are also subject to droughts when there may be no power available at all for extended periods.

Hydro power is also a built-in energy storage mechanism because you can make a reservoir from which you can draw water any time you want it. In fact, a reservoir is generally a good idea because it makes pressures constant. Of course, the big problem with hydro power is that you need to be by a river or stream (or dam) to be able to use it. Most people don't have this luxury, but for those who do, this is an excellent and economical source of solar power.

The good news is that you can get rebates and subsidies for wind and water power the exact same as for PV systems and water heating systems. They're all solar power and are therefore grouped together in this regard.

In the following sections, I explain the general concepts of using both wind and water power so that you can decide whether they're viable options for you, and if so, how best to proceed.

## Blowing with the wind

A wind turbine looks like a small airplane with a huge propeller — and that's because that's what it basically is. Wind pushes the vane (the opposite of what happens with an airplane propeller), which turns the propeller and forces the alternator to rotate, thereby outputting AC power.

Wind turbines (see Figure 14-2) work well under certain conditions, in which case the economics are good. They also have a number of operational drawbacks.

Here are the pluses of wind power:

✔ Power can be generated anytime, day or night.

✔ In some locations, wind is virtually a constant (magnitudes may vary, but output power is always available).

✔ Wind speeds vary over terrain, so you can find locations on your property that provide maximum potential. Ridge lines, coastlines, and the tops of barren hills are the best candidates.

✔ Wind is available almost everywhere, in all climates. In many of the worst climates, it's very powerful.

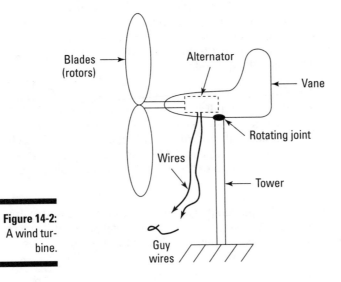

**Figure 14-2:**
A wind turbine.

And here are some drawbacks:

- ✔ Extremely high wind conditions may destroy or damage a unit. Turbines want a minimum amount of wind to begin working, but they also don't want to be subjected to tornadoes.

- ✔ Batteries are definitely required because you just never know when you'll get wind.

- ✔ In some communities, you may be prevented from installing a wind turbine due to noise and visual blight issues.

- ✔ Mounting is a bear. In general, the higher the unit, the better the output energy, all things being equal. One hundred feet is considered optimum, and that's a long way up there. The mounting must have a tremendous amount of integrity because of the torques that need to be withstood. And consider this: How do you raise the mount? You have to choose among a number of methods, each with pros and cons. And what if the mount were to break in high winds? Not only would this be very expensive, but imagine the unit crashing down onto something.

- ✔ Rotors are noisy and obtrusive when they're spinning (which they're hopefully doing quite a bit). The bigger the blades, the more power they'll generate, but they also make more noise because they're in contact with more air. They also attract a lot of visual attention and may clutter up a placid environment.

- ✔ Cost varies depending on how much wind is available. At some point, solar PV is a better investment. But if you get a lot of wind, wind is a cheaper and more reliable way to go than PV.

- ✔ Moving parts are never as reliable as stationary, so PV wins out in this category.

- ✔ Wind power is relatively obscure. Finding somebody to do your installation and servicing may be very difficult.

- ✔ Obstructions such as trees, houses, barns, and the like all affect wind speed. And because wind direction varies, obstructions in one direction may or may not be important in other directions. The best bet is to stay well away from all obstructions, and this may be prohibitive because of the topography or layout of your property.

In general, a wind turbine must be mounted at least 30 feet above the ground and at least 200 feet away from obstructions. There's 40 percent more wind at 100 feet than at ground level.

The smallest turbines (with 6-foot rotors) sell for around $1,000, not including installation (tower building and raising, plus wiring and so on). They output 400 watts of power at wind speeds of 28 miles per hour (this is a pretty good wind) and can withstand winds up to 110 miles per hour. You can use these numbers to get an idea of how many kilowatt-hours per day you can expect, on average; then you can devise a battery bank accordingly.

A turbine with a 15-foot rotor produces 3.2 kilowatts at 28 mile-per-hour winds for around $8,000, including installation. Fifteen feet is big, and the turbine makes a deep swooping noise. You'll know it's there.

Tower costs vary depending on height, but a typical tower kit for a small turbine runs around $700 (50' height). The installation is the tricky part. You better know what you're doing or somebody could literally be killed. The higher up, the more effective; but the higher up, the more expensive and dangerous the tower.

If you'd like more information in whether wind energy could be a viable option for you, check out *Wind Power Your Home For Dummies* (Wiley) by Ian Woofender.

## Using water resources for your power needs

Eight percent of United States' electricity comes from hydro generators located on major riverways. Water power is by far the biggest solar energy resource, and it can also be generated on small scales. Of course, you need access to a waterway, though either moving or stationary water can work. If the resource is stationary, you need to locate a hydro turbine somewhere below the top of the water, the lower the better because you get more head pressure. If the water is moving, you can simply dip a hydro turbine down into the water, at any depth you want.

You can use the movement of tides to generate electricity, but this is trickier to exploit than a river resource, although it has the potential to be more consistent over the course of a long time. However, you can generate electricity only when the tide is moving (either in or out, it doesn't matter which direction). See my book *Alternative Energy For Dummies* (Wiley) for more detail.

Hydro power has a lot of benefits if you're close to a suitable water resource. However, it also has drawbacks to consider. Here are some of the advantages of water power:

- ✔ You can potentially generate more kWhs per cost than any other energy resource, in particular solar PV panels.

- ✔ No batteries required (although they do make the system work better). You can assume your hydro generator output will be pretty constant, at least from hour to hour. Over the course of a year, you may have major variations.

- ✔ You can install a system of virtually any size power output if your water source is big enough, which is usually a requirement for economically viable applications. If your water source is small, power outputs will fluctuate quite a bit and may be zero for extended periods.

✔ You can generate power day or night, in any weather (freezing may cause problems, but you can usually design around them).

✔ Hydro systems have very long lives, are relatively trouble free, and require very little maintenance.

✔ A submersible hydro-power generator (moving water) for $1,200 gets you 2.4 kWh daily with a 9 mph stream. This is good economics, compared to other solar resources.

And here are some disadvantages of water power:

✔ Complex electrical system designs and mounting schemes are difficult with water pressures pushing all the time. This is not a trivial system design to tackle, although do-it-yourselfers can safely do it if they're patient and willing to try things, then adjust, then try, and then adjust until they've finalized the best arrangement.

✔ Waterways can dry up in droughts.

✔ Upfront costs are high, particularly for stationary water systems. In addition to the hydro generator, you need a good inverter to convert the raw voltages from the alternator into the standard household voltages that will run your equipment.

✔ Not many hydro generators are sold because so few people have access to a good water supply.

If you're interested in moving forward, you need to take some measurements based on your water source:

✔ **For moving water:** Measure the water flow and speed. *Flow* is how much water passes a given point in a minute. Generally, you won't be able to use the entire flow from a creek or river because that disrupts all wildlife functions. You can measure speed by simply tossing a stick out into the water and timing its travel over a given distance. (Measure the distance by pacing it out, one pace is around a yard.) An approximate measure is good enough, because speed changes so much that an accurate measurement gives no better results.

✔ **For stationary water:** You need to devise a pipe system that produces maximum pressure. This is a complicated (albeit interesting) subject, and you'll need to find some detailed resources. There are two types: high fall (head) – low volume, and high volume – low-head. The former is the type of system described previously.

You need water rights (legal classifications), which can sometimes be confusing and contentious. Just because you're next to a river or lake doesn't mean you have the right to use the water for generating power.

Here are the government agencies that control water rights:

- U.S. Geological Survey: www.usgs.gov
- Army Corps of Engineers: www.usace.army.mil
- Department of Agriculture: www.usda.gov
- County engineer: Do an online search or check in the phone book under county government agencies

You may have to deal with all four agencies before you start generating hydro power. Good luck!

# Chapter 15

# Glass Houses (and Plastic, Too): Greenhouses and Sunrooms

. . . . . . . . . . . . . . . . . . . . . . . . . . . . . . . . . . . . . . . . . . . . . . . . . . . .

## In This Chapter

▶ Deciding between the various solar room alternatives

▶ Exploiting the greenhouse and chimney effects

▶ Looking at some project types

▶ Arranging your solar room designs to maximize efficiency and effect

. . . . . . . . . . . . . . . . . . . . . . . . . . . . . . . . . . . . . . . . . . . . . . . . . . . .

The two types of solar rooms are greenhouses and sunrooms. *Greenhouses,* which are either connected to your house or separate, can help warm your house as well as provide inexpensive, delicious, highly nutritious food or beautiful, soothing, decorative plants. *Sunrooms,* which are always part of your house, add living space and square footage to your home for a relatively low cost, increase the efficiency of your heating and cooling efforts, and brighten up the ambience and decor.

Most home additions cost more than your property value increases. But with solar rooms, the greater design freedom and fewer building restrictions allow you to enjoy immediately profitable appreciations, especially if you do it yourself. You can get away with simple design elements that would never pass muster in your house proper, which means designs can be imaginative and nonconforming. And you can use inexpensive materials and support schemes that wouldn't work with the rest of your house. If you're clever enough, and patient, you can build a very nice solar room for next to nothing, which you can't say about any other room addition.

Every house is different, so no general design works universally. But all effective solar rooms follow general principles that can help you get the most efficient performance out of your new additions. If you've always wanted to make a big do-it-yourself impact on the basic form and structure of your house, this chapter tells you how to shine.

# Choosing a Type of Solar Room

The most basic difference between sunrooms and greenhouses is their function, namely whether you want to add comfortable living space for yourself or your plants. The following sections expand on this idea, noting how function carries over to design.

Both greenhouses and sunrooms can increase a home's privacy by providing a buffer between the outside world and the interior spaces. An attached greenhouse can also dramatically improve the aesthetic attraction of a home, at a very low cost compared to other remodeling alternatives (see Figure 15-1).

**Figure 15-1:**
Remodeling with a greenhouse or sunroom can dramatically alter your home's visual appeal.

## Growing green things

Greenhouses, whose main purpose is to support plant life, are very functional. Here are some basic greenhouse requirements:

- ✔ They need to invite the maximum amount of sunlight, which generally requires a glass ceiling or sloping glass walls.

- ✔ Their temperatures need to be regulated (the difference between the hottest and coldest temperatures needs to be kept to a minimum, dictated by the type of plants).

- ✔ They need adequate ventilation to provide oxygen for the plants.

- ✔ They need a water supply.

- ✔ The floors need to be able to withstand water leaks and mud spillage.

Well-designed greenhouses in cold climates can support the growth of plants all winter long, even in freezing conditions. In fact, greenhouses often need cooling means such as vents or radiant barriers to keep from getting too hot. On the other hand, greenhouses need to be sealed in order to maximize warming in the winter, so the interior air can get stagnant or too humid for comfort. The bottom line is that greenhouses need constant management and manipulation of air movement and sun exposure, not to mention the work needed to grow healthy, vigorous plants. Greenhouses are also prone to getting very dirty, so they're more commonly separated from the house.

## Basking in the sunroom

Sunrooms are designed as extensions of your home's living space. You can put some plants inside if you want, or you can completely enclose them from the elements and put in carpeting and fine furniture.

Leaving the doors and windows open between your house and a sunroom can give a sense of increased floor space and size to your home. There's no cheaper way to increase the square footage of your home than by building a solar room off your family room. In my own house, we practically live in our solar room, and we eat dinner out there nearly every night. It's the one project that we all agree we couldn't live without.

A well-designed sunroom can provide up to 60 percent of a home's heating in the winter, depending on the amount of sunlight available. Even in very cold climates, a sunroom can work efficiently. And particularly in a cold climate, a sunroom can provide needed relief from the gray doldrums of winter.

Such rooms generally have watertight roofs, which are usually solid. Too much sunlight is uncomfortable, and ultimately, your sunroom needs to be comfortable and inviting. Sunrooms can contain skylights and vents, and both increase the livability factor. Sunrooms also exploit the greenhouse effect to great advantage, but without plant growth, so managing the air quality and keeping things clean are much easier. The goal is to provide a warming space in the winter and to enhance a house's spatial and light composition.

# Exploiting Natural Effects

As with all solar projects, exploiting and enhancing the natural laws of physics are the necessary starting points. The basic idea is to catch as much solar radiation as possible during the day and then to keep that energy inside and use it wisely at night.

## Understanding the greenhouse effect

Everybody's heard of the greenhouse effect (see Figure 15-2); people invoke it to explain global warming, which is one of the big reasons most people are interested in solar energy in the first place. In this case, you want the greenhouse effect to work to your benefit.

Greenhouses work by the fact that sunlight enters the enclosed space through the glazing (window) and then gets absorbed and stored as heat energy. The heat is then constrained to stay in the enclosed space via the same glazing's insulation properties, plus insulation in the walls. You can enhance the greenhouse effect by increasing the amount of radiation the windows allow to enter while maximizing the glass's insulation against heat.

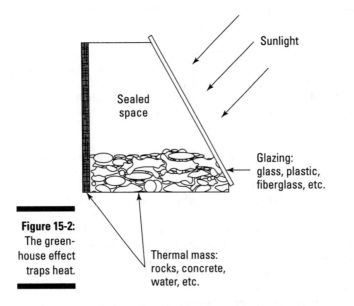

**Figure 15-2:**
The green-
house effect
traps heat.

Sunlight

Sealed
space

Glazing:
glass, plastic,
fiberglass, etc.

Thermal mass:
rocks, concrete,
water, etc.

The space must be well sealed to prevent leakage, although the greenhouse effect is often powerful enough to work well in relatively leaky environments. In fact, sometimes it works so well you can't tolerate all the heat.

## Up, up, and away with the chimney effect

When air heats, it becomes lighter in weight. Cold air weighs more, and hence gravity exerts a larger force on it. So the hot air rises, making the air in a room warmer near the ceiling than the floor. The phenomenon is called the *chimney effect* because of the way smoke goes up a chimney above a burning fire instead of spreading out into the room. Cold air is drawn into the fire, and the hot air rises quickly up the chimney. The air movement is brisk and consistent because the effect is pronounced.

With most solar rooms, the vents are movable, which means you can easily open or close them to take advantage of the chimney effect. When the solar room is warmer than the house, the vents open, allowing heat from the sunroom to enter the house (vice versa at night or on cold winter days). Or when the room's just too hot, you can completely close off the space from the rest of the house. Figure 15-3 shows how a sunroom works.

**Figure 15-3:**
Air warms in the sunroom, rises, and enters the house through the vent; inside, air cools, sinks, and then returns to the sunroom for reheating.

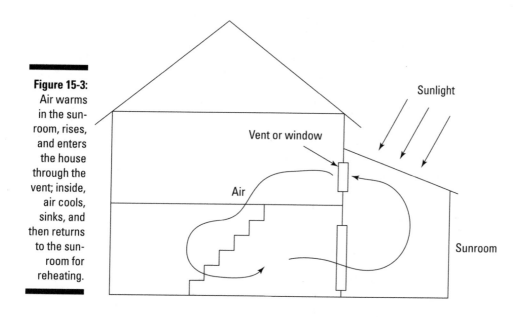

Sunlight

Vent or window

Air

Sunroom

Vent fans can enhance the chimney effect. Solar-powered ceiling fans work well because they don't need to be hard-wired (no code problems or expenses associated with electricians), plus they work hardest when the sun is hottest, which is usually what you want.

Solar-powered vent fans efficiently positioned right into the vent openings can also work to great advantage. Greenhouses can also be vented with solar-powered fans, which work well when the greenhouse isn't attached to the house and has no electricity available. Solar-powered fans are also better environmental solutions (see Chapter 13 for details).

# Taking a Quick Look at Typical Projects

As with all solar projects, kits are the best bet for the do-it-yourselfer, and a big industry is dedicated to manufacturing and selling prefabricated kits for costs anywhere between a few hundred dollars to hundreds of thousands. For example, small greenhouses made of plastic tarp (greenhouse film) and semicircular aluminum frame elements are great for growing food — no building code, no hassle. They're very cheap, starting at around $300. You can add options such as automatic vents, automatic watering systems, blinds and reflectors, and so on.

Extruded aluminum frame sunrooms are common, and their availability is proliferating as costs decrease. You can build them with all windows, or for better structural integrity and some relief from bright sunshine, with insulated fiberglass panels interspersed with windows. They usually have solid, insulated roofs to give the space more of an interior feel.

Prices start at around $10,000 for complete do-it-yourself kits and can go much higher. Installations aren't that expensive, so you should get a quote for doing it yourself versus contracting, factoring in how much work installing it yourself would take. Contractors come in and put these things up in a few short days, and they know the ins and outs of getting permits and the like (when working with contractors, building the project outside of the permit process is almost impossible — see Chapter 19). People who get these sunrooms are generally very happy with the result and recommend them to their friends and neighbors.

For the money, if you're a do-it-yourselfer, a custom design can add the most visual appeal to your home. But you can still get kits of windows and wall materials to make things easier. Look on the Internet and in the phone book for ideas. You can get books for greenhouses and sunrooms in big hardware stores or through bookstores, both tangible and online. If you're patient and spend enough time on research and design, you can do a very good job and you'll be happy with the result, both functionally and aesthetically.

# Getting the Most Out of Your Design

Solar rooms all have the same basic fundamental components. There are a million ways to build a solar room, but some practical generalizations can ensure a successful project. In this section, I explain the basic components of a solar room and give you some tips on building and design.

# Anatomy of a solar room

Solar rooms all have the same basic components. The following sections explain the functional elements of an ideal solar room (also see Figure 15-4).

The transparent cover (also referred to as glazing, or windows) allows for sunlight entry. The rock pile is thermal mass, which stores heat and serves to regulate a consistent temperature in the space. A reflective wall lining is optional; it serves to reflect sunlight down onto the thermal mass and plants. Insulated north wall is also optional; the insulation works with the thermal mass to keep heat in the space and maintain a consistent internal temperature.

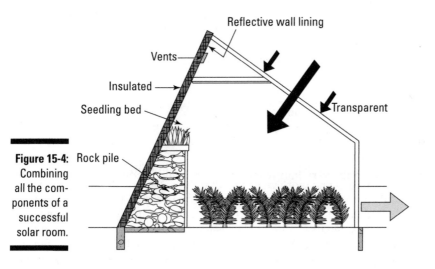

Reflective wall lining

Vents

Insulated

Seedling bed

**Figure 15-4:** Rock pile
Combining
all the com-
ponents of a
successful
solar room.

Transparent

## Letting some sun shine in

As with all solar systems, a collector, or *transparent cover,* allows sunlight to enter. The larger the area, the more energy captured. You can control seasonal and daily variation by orienting the collector's east/west direction (azimuth) and altitude (upward angle toward the sky). You can also maximize sunlight in the morning (good for comfort and generally good for sunrooms because you want a warm space early in the morning, and afternoon sunlight is often too harsh and hot for comfort) or in the heat of the afternoon (best for maximum heat capture). Chapter 5 explains more about available sunlight.

Commercial greenhouses are usually made with glass ceilings (roofs), but you can get the same open effect by angling the window glass (refer to Figure 15-4).

Double-pane glass works well as a collector, and a number of window coatings and other optical engineering methods can also work to good effect because they allow light to pass through but also insulate for heat. Some inexpensive plastic materials also work well, though they tend to blur the view, so they're generally more suitable for greenhouses than sunrooms.

Greenhouses often have solar reflectors mounted into their roofs in the summer because there's just plain too much sunlight. You can choose from a range of clever movable insulation methods, such as a blind mechanism loaded with radiant barrier reflective material (see Chapter 9). Hoods, overhangs, and awnings can also control the seasonal and daily variation of sunlight. Window blinds are often used in solar rooms to prevent heat escape at night, to keep the room from getting too hot in the summer, or to enhance privacy. They also make the room much more attractive.

### Keeping the heat

After the sunlight enters the space, an *absorber* captures the energy and transforms it into heat. Dark, rough surfaces — such as gravel floors in a greenhouse or dark furniture and carpet in a sunroom — work best.

Heat storage is accomplished via thermal mass and material (see Chapter 4). Rocks are cheap and work well, and water is a good choice because it's cheap, holds heat, and is readily available. For maximum functional effect, drums filled with water may be painted black and set in the direct sunlight. The heat storage means is almost always located directly below or behind the absorber, and in many instances, it can be the same thing. Figure 15-5 shows some heat storage methods. A *water wall* is a rectangular space filled with water; it's more common to see individual drums stacked on top of each other than an entire wall made to hold water; either option works well. The only thing that matters is how much mass is present. In either case, take care to close off the water to prevent evaporation and keep mosquitoes from laying eggs in the stagnant water.

Sunrooms don't require as much thermal mass because you can close them off from the house when needed. However, if you want to maintain maximum comfort around the clock, some form of heat storage is a good idea. Concrete floors are good solutions because they provide not only mass but also a good, solid underpinning to the room.

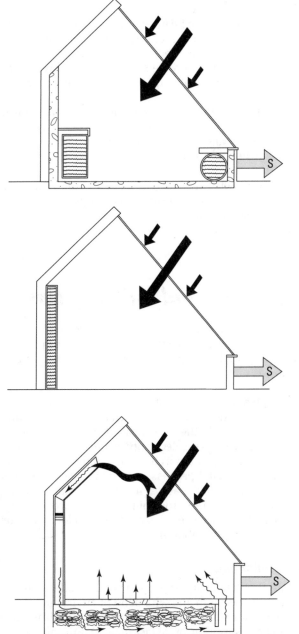

**Figure 15-5:**
Storing heat maintains consistent temperatures inside a solar room.

### Putting heat on the move

*Heat distribution systems* move the heat, as appropriate. The best methods exploit the chimney effect (the idea that heat rises), but fans and vents are also effective, if designed properly. Keep in mind that heat can be moved by conduction, convection, or radiation (see Chapter 4). Always work with the natural movement of air (see Chapter 13), especially your prevailing winds.

*Control mechanisms* decide when and how to move the heat. Fans may be switched, vents may be opened and closed, thermostats may switch off and on at preset temperatures, and so on. In most applications, you're the controller — you decide when to open and close windows and vents and blinds.

If you'd rather opt for non-electrical means, an automatic vent opener — a compact, mechanical, heat-activated device — can open windows, skylights, greenhouse vents, and so on at a temperature you select. Because they don't use electricity, you can mount them anywhere. For heavy windows and vents, simply use multiple units.

## Following design and building tips

A do-it-yourselfer can design and build a solar room for a relatively low cost, with relatively low risk. You don't need electrical wiring or plumbing, and you don't need to obey all the building code requirements that are unavoidable inside of a house.

You can work through a project without having to apply for building permits or to have the property value reappraised (with the commensurate increase in property taxes). Simply build the room against the house, but don't actually connect it to the house proper (no electricity, water, or continuous wall connections) — in this case, it's just a porch. Keep in mind, however, that when you don't go through the permit process, you can't advertise your solar room as part of the square footage of the actual house. This may or may not matter to potential buyers, who'll appraise the value on their own terms when they take a tour with their Realtor. To be sure, check with a local Realtor and get his or her opinion of your local situation.

### Location and surroundings

Before you get into the details of design, you should make some decisions about where you want the room and what kind of environment you want to create. Always build your sunroom on a southern front (although eastern sunrooms are popular as well for breakfast niches). Put the absorbers and thermal mass on the north side, at the back of the room. If you can't, your room will still work but not to maximum advantage.

The best location in your house is adjacent to the kitchen for attached greenhouses and adjacent to the living room or family room for sunrooms. These locations afford not only the most efficient use but also the most use in general.

Try to use as much of your home's existing walls as possible, preferably on both the east-and-west end of the sunroom. These walls are already insulated, and they match the house.

An easy and straightforward candidate for solar rooms are existing porches and decks that already have the basic support structures and flooring in place. All you need to do is build up and around.

As for surroundings, you may want to plant deciduous trees at the same time as the building of your solar room. Sometimes solar rooms can get too hot in the summer. In particular, western trees block the harsh late afternoon sun. You can cut back the sun in the summer and allow it all in during the winter. Plus the room will be much prettier if you can look out at a nice tree. Partnering with Mother Nature gives a certain continuity to the entire project. (See Chapter 8 for more details on how to plan your landscaping.)

### Materials and design

Kits are widely available for both greenhouses and sunrooms. You can find showrooms where samples of the final product are available, so try to touch things before you buy; photos are often misleading and rarely reflect reality the way you expect.

Of course, you may want to design your own solar room. As much as a well-designed solar room can enhance your house's aesthetics, a poorly designed one can make your house look awkward and uninviting. Make very good drawings done to scale before you begin building (use graph paper and let 1 inch of graph paper represent 1 foot). Consider all the angles, and if you can, draw some different perspectives (from the street, for example). The more thought you give to the appearance before you start, the better the odds that your solar room will increase the value of your home.

As you draw your design and try to figure out what materials you want to use for your glazing, remember these tips:

- ✔ Sloped glazing allows for more sunlight entry, but it also gets much dirtier and is leakier and harder to install.

  The do-it-yourselfer should avoid horizontal glass because it can be dangerous if it breaks and therefore takes special engineering techniques to ensure integrity.

- ✔ In cold climates, use between 0.65 and 1.5 square feet of double-pane glass for each square foot of building floor area. In temperate climates, use between 0.3 to 0.9 square feet for each square foot of building floor area.

In practical terms, don't worry if you can't achieve these ratios. Anything will work to your advantage. Solar rooms are usually compromises between the best physics and the best aesthetics and cost. In general, aesthetics should win for sunrooms, functionality should win for greenhouses.

✔ Glass shops often do retrofits of entire houses; they remove the old windows and install new ones. You can buy the old windows for next to nothing (they're usually single pane, but if they're almost free, you can't complain). Reusing such glass may lend a discontinuous visual effect, so be prepared to add some paint or other finishes.

✔ A lot of new plastic materials are very effective at glazing. These materials are unsuitable in your house, but they're okay in a solar room. If you're building a strictly functional greenhouse, you can use corrugated plastic roofing panels on both the roof and sides. These make for very easy designs and construction because you can nail everything up very easily, plus it comes with foam sealant strips that are very effective and cheap. You can use the clear stuff — or the somewhat transparent green stuff if your climate is too sunny and hot for full exposure. Greenhouse kits made of plastic sheets instead of windows are also easy, cheap, and effective. But plastic doesn't last long, and you'll be required to replace it at some point.

If possible, use the same roofing and siding and window materials that your house is made of for better aesthetics. If you can't, don't try to match at all — just do something that's complementary but totally disparate. Nothing mismatches more than an attempt at matching that doesn't quite cut it. This is like matching clothes; blue and orange match up much better than orange and a little bit different orange, which just ends up looking cheap.

Especially in a greenhouse, make sure that you pay attention to ventilation. Chapter 13 can tell you more about fans and ventilation schemes.

### Cleanliness

You may be able to tolerate dirty windows in the rest of your house, but solar rooms have a tendency to magnify imperfections. Let the sun shine in!

Plan on devising a good, efficient, easy method of washing the windows often. Keep good-quality squeegees on hand (get long poles), and allow for drainage of ammoniated chemicals below (don't put sensitive plants directly beneath the windows because they'll get the chemicals into their root systems). Don't wash windows with a hose attachment; otherwise, you'll end up with water sediment building up and streaking.

# Part IV
# Exploring Full-Scale Photovoltaic Systems

"I think our solar panels are drawing in too much energy. I just plugged in my hair dryer, and I swear the sun dimmed."

## In this part . . .

**M**aking the decision to install a big PV system on your roof is important because it involves a lot of money and has to be done just right. In this part, I describe the equipment that's involved and explain how to choose the best equipment for your particular application. I show you how to calculate the finances, find a contractor, and come to an agreement with your contractor on the size and location of your PV system. Finally, for those of you who want to completely cut off all ties to the utility company, you can try off-grid living. I show you how to calculate the best system for your needs.

# Chapter 16

# Taking a Close Look at Photovoltaic Systems

## In This Chapter

▶ Understanding the benefits of large-scale PV systems

▶ Getting into the hairy details of solar equipment

▶ Contracting and getting the job done properly: The process in a nutshell

*I*nstalling a full-scale PV system at your home is the king of solar investments, and it's the solar option that many people are most interested in. You can offset your entire electric bill, at essentially zero pollution (aside from the pollution that is emitted during the manufacture of the solar equipment). The net cost is high, ranging from more than $8,000 to up to $100,000. For some, it's a simple decision because the main goal is to eliminate pollution; for others, payback is the most important consideration.

There are three basic types of PV systems: *intertie*, which work in conjunction with the utility grid: *standalone*, which require batteries and other special support hardware and are usually installed where grid power is either nonexistent or very expensive; and *hybrid systems*, which are both intertied to the grid and have battery backup.

In this chapter, I give you broad picture of full-scale PV intertie systems. I restrict the discussion in this chapter to intertie systems because they offer the best return on investment, comprise the vast majority of PV installations, and make the most economic and environmental sense. You can find out more about standalone systems in Chapter 18. Even for those interested only in standalone, I recommend a good understanding of the info in this chapter. In the Chapter 17, I explain how to determine whether such a system is right for you.

# The Benefits of Powering Your Home with PV Systems

Most PV customers are interested in solar powering their homes because they realize that solar power is good for the environment, among other things (Chapter 1 describes the various pros and cons of going solar). Few other investments can positively impact the environment as much as an intertie (utility-connected) PV system. But in many cases, a PV system is also a better financial investment than most other investment options, and the vast majority of PV customers want to know, in precise detail, what they can expect their new system to produce. For a given house, installing the right system versus the wrong one can make the difference between a very good investment and a poor one. Here's a short list of PV system advantages:

- ✔ **Lower utility bills:** The most obvious advantage of installing solar PV is that you'll reduce your electric utility bills. It's easy to determine how much your system will produce, in terms of energy. It's a lot more difficult to predict how this will affect your monthly utility bill, and that's the bottom line.

- ✔ **Simplicity:** The system architecture of a PV system, while considered high-tech, is very simple. There are no moving parts. The simplistic nature of PV systems means

  - **Maintaining a PV system is simple.** You usually don't need to do anything at all. While it may seem like your panels are getting very dusty, the fact is, the affect on power output is much less than you'd think.

  - **Simplicity ensures reliability.** The systems have long life-times because there are no moving parts. You can expect your solar PV system to operate for more than 25 years.

  In fact, no other alternative energy option affords the simplicity of a PV system.

- ✔ **Subsidies:** Government has a vested interest in promoting solar power because of its overwhelmingly positive environmental impact, as well as the desirable political goal of energy independence. The best way to promote solar power, of course, is to get more people to invest in it, and the best way to do that is to subsidize it and drive the net costs down. (*Net cost* is the amount you pay, out of pocket, for your system; it's the starting price minus all the subsidies and rebates that you receive.)

  Subsidies, rebates, and tax breaks are widespread, and they're becoming more prevalent every day. In some parts of the country, the total discount on a system cost can be more than 50 percent. Pay too many taxes? Why not get something back from the government for a change? For more on subsidies, see Chapter 20.

## Buy now or wait? Timing your purchase

So should you buy now or wait? Although solar contractors inevitably tell you to buy now, the fact is, nobody knows where net costs are going. The only certainty is that utility power is going to become more expensive, and probably much more expensive. But because government incentives change so much, it's impossible to predict what the net cost of a system is going to do simply because it's impossible to predict what politicians are going to do. Net prices have come down in the last few years, particularly in light of the ITC (Investment Tax Credit) change (I explain this in Chapter 20).

People who bought systems a few years ago have seen their neighbors install identical systems for thousands of dollars less.

Ultimately, you'll have to answer the timing question for yourself. Stay abreast of the changing laws by reading newspapers and magazines. You'll find literally thousands of alternative energy promotions being discussed by various government bodies, and any one of these can impact the net cost of a solar PV system.

✔ **Pollution mitigation:** You can literally erase more than 40,000 pounds of emitted carbon dioxide per year by going with a large-scale PV system. And if you're hooked up to the utility's power grid, whenever you're not using the power you generate, someone else *is*, so you're offsetting another person's carbon footprint as well.

✔ **Hedging:** All investments require you to predict the future. If you think energy costs are going to rise quite a bit, solar PV is a very wise option. When you bring your power bills down to an average of zero, they'll stay there regardless of how much energy costs rise.

✔ **Lifestyle insurance:** When energy costs rise, people with solar power will still be able to use the same amount of energy they were using previously. Those without solar will be scrambling to conserve, or scrambling to buy solar (which makes your equipment more valuable simply because demand has increased).

✔ **Appreciation:** Your property value can increase by more than the cost of your original PV investment — without raising your property taxes (see Chapter 6)!

✔ **Intangibles:** PV customers enjoy many intangible benefits. It's fun to see your power meter go backward (take that, utility company!). Solar power is cool. When you put panels on your roof, people notice, and they'll be nicer to you (well, maybe). Ask people who have installed solar PV and they'll very likely tell you that it's great – they're very happy they took the plunge, and they'll advise you to do the same. They'll tell you that they're saving money, but there's always an undercurrent of personal satisfaction that seems more important than the financial rewards. When you install solar PV, you have earned the right to swagger.

# *Looking at the Basic Components of a Intertie System*

Solar intertie PV systems are not particularly complex. They're expensive, but in terms of high-tech electronics, they're pretty simple. First there are panels, which collect the sunlight and turn it into electricity. The DC signals are fed into an inverter, which converts the DC into grid-compatible AC power (which is what you use in your home). Various switch boxes are included for safety reasons, and the whole thing is connected via wires and conduit. The solar PV-generated power is connected to your home's grid at your main fuse box (see Figure 16-1),

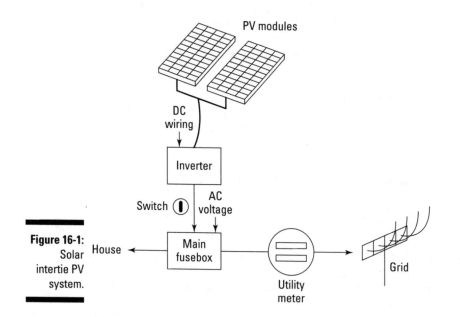

**Figure 16-1:**
Solar intertie PV system.

Keeping it simple pays. In general, the vast majority of customers install the simplest possible system on their roofs because this allows for the best return on investment. You can get real fancy with solar PV, but costs rise fast. You can choose to put a "skirt" around your solar panels, for example, to hide the underlying mounting frames and improve the appearance, but you pay extra for this option, and the production suffers as well because the panels operate at a hotter temperature (the skirt prevents cooling breezes). My advice? Stick with proven, field-tested equipment that's as simple as your situation merits and you'll achieve the best return on investment.

# Various types of panels

PV panels, which cost anywhere between $2.40 per watt to over $5 per watt, are the single biggest expense of a PV system. Their placement and mounting affect your system performance more than any other facet of the job.

Different types of panels are finding their way onto the market. The configurations of these different types of panels (how they're combined physically) govern how much space they take:

- **Rectangular:** The most common type of panel is a rectangular, aluminum framed complex of individual solar cells (see Chapter 4 for more details).

- **Triangular:** Triangular configurations can match the contour of your roof and offer a pleasing, symmetrical appearance from street level. Expect to pay more for these.

- **Integrated:** These can be mounted directly over tiled roofs and they match the undulating surface, making them great for Spanish-style roof. Expect to pay a lot more for these. You can also "integrate" panels directly into your existing roof, matching the roofs tile shape and size. These are expensive, but offer a very pleasing appearance.

- **Flexible:** These panels come like a big roll of tape and can be installed on flat roofs very easily. This type of panel is not as efficient, so it takes up more roof space per watt. But the price per watt is lower, so there's a tradeoff.

Solar panels are either bluish black or solid black; (color doesn't matter in terms of performance) and the aluminum frames are either anodized clear or black (black is the more popular option these days).

Many customers are concerned with a panel's efficiency, which is the ratio of power output to square footage ( the latter is often referred to as *footprint*). The fact is, efficient panels usually cost more per watt, and if you have enough roof space you don't need to maximize panel efficiency. If you're pinched for roof space, you may need to invest the extra money for efficient panels.

## Warranties

Most PV panels are warranted for 25 years, but beware; panels degrade over time (just like everything else, they get worn out). System warranties specify a percentage of original power output over time, say 80 percent after 25 years. Over time, your system puts out less and less energy; it's inevitable.

If your manufacturer has gone out of business, you may have a very difficult time getting anyone to honor the warranty. First, you may not have a warranty at all, and second, you may be stuck trying to find a replacement panel from a different manufacturer that's got the same size and performance as your original, which may be impossible. At the very least, it's likely that a different panel will look different, and stick out like a sore thumb. If you put a mismatched panel into an array, the overall system performance may suffer. The solution? Stick with reputable manufacturers who have been in the business for some time; they're more likely to be in business in the future.

### Manufacturers

Stick with a reputable manufacturer that can document a proven track record. There are all kinds of new panels and manufacturers, particularly from Asia. Many customers, in my experience, shy away from Chinese panels due to the fact that governmental regulation is so lax there. We've all read stories of Chinese products that have turned out to be dangerous because noxious chemicals were used, inadvertently or not. The panels I have seen from China work fine, and the quality is good, so don't automatically discount them. On the other hand, some skepticism is in order.

Check the quality of panels by reviewing information on the Internet. One leading solar installation contractor in the United States had to recall all the panels the company installed over a two-year period because a defect was discovered. The company made good on its warranties, but customers went without energy production while the panels were being changed.

### Panel maintenance

Many new customers feel the need to clean their panels, but it's usually not worth the work, which can be difficult and dangerous, especially if they're mounted on the roof. If the panels are ground mounted, cleaning them is usually easy. Simply hosing the panels off with water is fine— unless you have well water, because the sediments calcify on the panel face and over time a layer of crud builds up and affects your system performance. Plus the layer of crud is extremely difficult to get off and it looks cruddy.

Don't rub the panels with an abrasive brush or sponge. You'll scratch the surface, and this will result in degraded performance. The coatings on panels are extremely important; they ensure long lifetimes and good performance.

## Mounting equipment

Mounting your PV panels is of critical importance. First, you need to mount the panels where they'll get maximum sunshine over the course of a year (see Chapter 5). But the more difficult problem is to mount them with enough integrity that they'll stay put for 25 years or more.

PV modules are temperature sensitive. When the weather gets hot, power output goes down. Systems often output more power on a cold, clear spring day than on a hot, muggy summer day, despite the fact that there's a lot more sunshine on a summer day. Because the modules are temperature sensitive, you need to pay attention to how they are mounted.

The following sections outline your mounting options.

### Roof-mounted systems

Roof mount PV is by far the most common and comprises around 95 percent of all residential PV systems. Mounting panels onto your roof is the least expensive way to install solar panels. It also raises the panels above ground level so that shade issues, like trees and neighbor's houses, are minimized. Four types of mounting systems are in common use:

- ✔ **Rack mount:** The PV panels are captured by a metal framework specially designed to allow easy attachment and disattachment of the panels. The panels are almost always parallel to the roof surface. A wide range of different types of racks are available. In many cases, a particular panel manufacturer will also provide the rack mount specifically for their panels.

  The most important thing about a roof mounted rack is that it doesn't introduce leaks in your roof. Make sure to come to an understanding with your contractor about this issue. A good contractor knows how to install racks without introducing leaks.

- ✔ **Stand-off mount:** The panels are supported by a frame built above the roof. Unlike the standard rack mount, the angles can be adjusted so that the panels aren't parallel to the roof plane. Stand-off mounts are ugly, but if you don't care about the visual appearance, you can get better production. These types of mounts are generally used for homes that have way too much shade on the southern roof exposure and a sunny northern exposure.

- ✔ **Direct mount:** The panels are attached directly to the roof. There is no air gap between the roof and the panels, so there is no cooling. While you save money on the cost of the rack equipment, your system production suffers.

- ✔ **Integrated mount:** PV panels replace conventional roofing materials and attach directly to the roof's rafters. These are sometimes referred to as *BIPV*, or *Building Integrated PV*. You save on the cost of roofing materials, and the visual appearance is nice. If you're retrofitting an existing building, it's not worth it to use integrated mount (because of the cost and waste of removing the existing roofing materials and throwing them away). However, if you're building a new home, BIPV may be the best option.

### Ground-mounted systems

If you have enough space available, you can mount your panels in a specially engineered rack structure affixed to the ground. In other cases, roofs are simply too complex, with too many vents and odd angles, to support a large array of solar panels, so ground mount may be the only option.

Ground-mounted systems offer both pros and cons: On the pro side, you can orient the panels directly south, at the optimum tilt angle. This ensures the maximum amount of production over the course of a year. You also won't have to worry about leaks in your roof, and the panels are easier to maintain and change, if the need arises.

On the negative side, ground-mounted systems are more expensive than roof-mount because they require concrete posts and rigid frames. Wind is also more of a problem with ground mount. Ground mount systems require suitable land space, and the appearance is definitely industrial. You'll be putting a big, ugly, visible array somewhere on your property.

### Installation considerations

How much you pay for a roof-mounted system depends on the type of roof you have:

- ✔ **Composition roofs:** These are the least expensive type to mount to.
- ✔ **Flat concrete tiles:** These require more work on the part of the installers, and you'll pay around 5 percent more for a flat tile roof.
- ✔ **Spanish tile:** These roofs are the most expensive because these tiles can break and shatter if they're not treated with kid gloves. Spanish tile roofs are also prone to leaking, if the installation is not done just right.

The best bet, as far as keeping the panels cool, is to mount them where there are consistent breezes (although this may be at odds with facing them to the south). Panels mounted around 6 inches above the roof cool down better than those mounted directly onto the roof. But this 6-inch height can be ugly; you don't want to see the mounting hardware beneath the panels because it ends up looking "industrial." So there's a tradeoff, and the typical mounting distance is around three inches.

Skirt racks are comprised of a surrounding skirt that hides the underlying rack mount hardware, but this prevents breezes altogether and results in lower energy yields. But these skirts do look very nice. If appearance is important, consider a skirt system, but you may end up having to install a larger system (more panels) to get the energy output you're looking for. The return on investment decreases with skirts but the sex appeal increases.

# DC-to-AC inverters

Inverters are the next-biggest expense after PV panels. Inverters take the low-voltage, high-current signals from the PV panels and convert them into 120VAC (or 240 VAC), which is directly compatible with grid power. Inverters cost around $0.70 per watt, or around $2,600 for a typical application. From a reliability standpoint, they are generally the weak link in any PV system, so quality is a must.

Most installations use only one inverter, but for big systems, having several inverters is common. You can install an inverter that's larger than the power output of the array you're installing, and then at a later time install more panels.

TIP

Consider using two 3kW inverters instead of a single 6kW inverter because if one breaks, you still have half your capacity. If the price is the same for two half-sized units, compared to one full size unit, it's a good idea to consider the two smaller units. You can potentially get better shade performance as well.

Manufacturers use different technologies to build inverters. Early models used crude transistor switching circuits to create "square waves," which were then fed into a transformer to increase the voltages to those useable in a home. The resulting outputs were very noisy and efficiency was poor.

With the advent of sophisticated integrated circuits and high-power field effect transistors (FET's), new inverters output sine waves, which is what the utility feeds into your home electrical system. Efficiencies are much better because the circuitry matches grid power to a higher degree. Whenever your inverter output mismatches grid power, you lose money.

## Types of inverters

There are two basic types of inverters in use for residential applications: string and microinverters.

### String inverters

String inverters are large boxes mounted near your power meter or main fuse box. In the vast majority of applications, only one inverter is used. In large systems, two or more inverters may be used, and they are simply connected in parallel.

It's of critical importance to design the "strings" of PV panels that feed into an inverter, a job that's almost always better left to the experts. For instance, an array consisting of 24 panels may have four series strings of six panels connected in parallel, or six series strings of four panels connected in parallel. The configuration makes a difference in terms of system energy output, and it depends on how much shade is present. There may be numbers of panels that are simply not doable, like 19. Since this is a prime number, the panels can't be arranged in series and parallel strings (the only possibility is a string of 19 series panels, which may overload the inverter). So be aware that you can't simply choose any number of panels. Similarly, it's best to use identical panels in order to harmonize string outputs, which maximizes over-all system efficiency. You can't combine a hodgepodge of different modules with any degree of system efficiency.

String inverters are the least costly option for the amount of energy a system can output and the technology has been proven over years of widespread use. You also have many different manufacturers to choose from, most of them very good, with excellent track records. Plus, string inverters are easy to maintain because they're mounted in accessible locations.

String inverters aren't without their disadvantages, which include the following:

✔ Shading becomes a problem because arrays are arranged into series and parallel "strings." If a single panel is shaded, the entire string suffers, and the total energy output of the inverter suffers as well.

✔ When a single panel goes down (breaks), system output is cut consider-ably. How degraded the output is depends on how the individual panels are configured into strings.

### Microinverters

Microinverters are the new fad. Each PV panel is outfitted with its own inverter, and the output of each is AC (which matches grid power). The number of microinverters matches the number of PV panels. In the future, it's likely that the vast majority of systems will use microinverters, but for now they're relatively new and the reliability risk is high.

Each microinverter outputs its own AC power, and all the microinverters must run in sync with each other (meaning their AC output signals must match up in both voltage and waveform). This is a very difficult electrical engineering problem, and requires that a special electronic box be mounted near the panel array. Each microinverter feeds its signal into this box, where all the signals are combined and then the grand total is fed down to the main fuse box, or power meter.

With microinverters, there's no need for complex string-sizing calculations, so the design is much simpler. Plus, there are no high DC voltages to contend with. You can also mix and match panel types.

Another benefit is that the efficiency of the overall system is not affected nearly as much by shade. If a single panel is shaded, all the rest of the panels continue to output their maximum amount of power. If you have serious shading issues to contend with, microinverters are the way to go. Similarly, if a single panel goes down, you only lose power output from that single panel.

Still, microinverter technology is brand new and unproven, which counts for a lot, since you'll be expecting upwards of twenty five years of reliable performance from your PV system. In addition, these inverters are more expensive, per system output power. Expect on the order of $0.40 more per watt. In a 4,000 watt system this is $1,600. (Keep in mind, though, that if you have shade, you may recoup this extra cost in terms of better system performance over the course of the year.)

### Inverter specifications

Pay attention to these specifications:

- ✔ **CEC Rated Power Output:** This tells you the maximum output watts from the inverter and varies over operating temperature.

- ✔ **Maximum recommended PV input power:** This must never be exceeded by the power output from the combined panels or the output becomes *clipped,* which means that your inverter runs inefficiently. In other words, you must match the inverter to the panel array properly. Use PV Watts (enter "PV Watts" into your search engine) to find out whether your inverter is large enough for the array you're considering.

- ✔ **Maximum open circuit voltage:** The PV array's maximum open circuit voltage must always be less than the inverter's limit or damage may occur. Calculating this number for a given array is a complex engineering problem.

- ✔ **PV Start Voltage:** This tells you when the inverter will begin to operate. In the morning, when the sun comes up, the PV panels begin to output power, but inverters require a minimum amount before they start outputting their own power into the grid. This is an important specification because it relates to the overall efficiency of a system. You want your system to run as long as possible over the course of a day.

- ✔ **Maximum Power Point Tracking (MPPT) Range:** MPPT circuits enable the inverter to harvest the most amount of energy over the course of a day.

### Inverter features

Efficiency is the most important factor when evaluating inverters. Unfortunately, individual efficiency measurements can be misleading. Manufacturers commonly cite peak efficiency, for example, but this only occurs when the inverter is fully loaded and experiencing maximum input power from the PV panels.

When choosing an inverter, look for one that has a high efficiency over a broad range of conditions, including input power, temperature, output load, and varying grid power characteristics. Some inverters claim high efficiency, but the efficiency relates to one number only, say peak efficiency. You need to consider the efficiency over the entire workload.

The following sections outline other things you should look for in an inverter.

If you're going to DIY, get a copy of the inverter's instruction manual prior to selecting that inverter. We've all seen garbled translations of foreign languages, and if you can't understand what in the world the instruction manual is talking about, you may have difficulty installing the inverter correctly, which can affect safety as well as efficiency.

### Modular circuitry

Ease of servicing is important because problems do occur. Modular circuitry can be replaced in the field, at low cost. An inverter that needs to go back to the factory costs you in terms of down time, which means that you won't be reaping the benefit of your solar PV system when it's down.

### AC and DC disconnect switches

The trend is to include these in the inverter, saving the need to install external switching means and junction boxes and the required conduit. The total system price will be less, and the visual appearance is cleaner as well.

### Monitoring functions

These aren't strictly necessary because they don't increase production (which means your return on investment suffers because you pay more for the same energy output), but they can tell you how your system is working (or not working, as the case may be).

The simplest monitor function is a display on the inverter that shows how much power is being produced at any given time (in kilo-watts). You can see when the system kicks on in the morning and when it gives up the ghost at the end of the afternoon. You can see how the power output varies on hazy or partly cloudy days. You can see how the power output suffers when it's extremely hot out. You learn your system's personality traits.

More advanced monitors yield a reading of total generated energy, either weekly, daily, or lifetime. Some tell you how much greenhouse gas you're offsetting (avoiding as emission into the environment).

The most popular monitor is a digital remote station that you can put inside your house so that you can read it without going into the garage (or wherever the inverter is installed). These cost anywhere between $250 up to $1,000, depending on the features, and how many inverters you may have. Many contractors include these with every system they sell, and the price is often negotiable.

You can get a monitor that connects to your computer so that you can call up on the Internet, even when you're at work or on vacation, to find out how your system is working. These latter systems also provide you with an opportunity to do all sorts of complicated calculations. With these devices, you can also have your contractor access the data and decide if your system is working the way it's supposed to. Many contractors offer to do this for free; ask them.

Once your contractor has given you a bid for a job, pretend that you're on the fence between choosing him and somebody else, and then ask if he'll throw in a free monitor to cement the deal. He probably will.

### Safety features

These include switches and fuses that prevent dangerous situations. The inverter should also show, on some kind of display, what the fault is in case of a safety mishap. For example, "OV" may indicate an overvoltage from the grid.

Protection circuits prevent the inverter from being damaged if something goes wrong. For example, if there is a high voltage surge in the grid power, the inverter should be able to deal with this without permanent damage.

### Expandability ports

These ports are often wise. You can purchase a larger capacity inverter than what you currently plan on installing (in terms of PV panel capacity) so that later on, in a few years, you can add PV modules. With expandability ports, making the transition to a larger system is easy. Without, you may have to reconfigure your PV panel array, which could be very costly in terms of rewiring and recalculating the strings of panels.

### Ability to output energy over wide range of conditions

Over the course of a day, sunlight levels change quite a bit, so the operating ranges of an inverter need to be very wide. A good inverter can efficiently output energy over a wide range of conditions. What you're after with an inverter is the maximum energy output over the course of a day.

Given the same PV array and the same exact sunlight and temperature conditions, different inverters output different amounts of energy (kWhs) over the course of a day. The inverter that outputs the most energy is the most efficient, but the way specifications are written for inverters it's nearly impossible to tell which one is truly the best. Once again, consult consumer guides.

### Cooling options

Inverters come with either passive or active cooling. Active cooling works best (an internal fan moves air over the parts), but whenever moving parts are used, reliability inevitably suffers because moving parts simply break more. On the other hand, passive cooling requires large, heavy heat sinks mounted in locations where airflow is easy and direct. Passive costs less, but may not produce as much energy since the electronics may be operating at a higher temperature. If you've got good breezes and shading where your inverter will be mounted, opt for passive.

### Inverter warranties

Warranties are extremely important. Most last from 5 to 10 years, but you can find 15 year warranties. Read the fine print. Does the warranty include labor or just parts? Does the warranty include shipping and handling? And who provides the warranty: the manufacturer or the contractor? Some contractors even warrant for lost production; in other words, they pay you for the cost savings you aren't achieving when your unit is down.

You can often purchase an extended warranty, but these aren't usually worth the extra cost.

### Mounting options for the inverter

Different mounting options for the inverter are also worth considering:

- ✔ **Directly next to the main fuse box, on the outside wall of your house.** This is the most common mounting location for an inverter. They're made to spend their lives outdoors. The only real problem they may have in this location is when they're in direct sunlight and end up getting too hot.

- ✔ **Inside your garage:** If your main fuse box is on a southern front with a lot of direct afternoon sunshine, you may need to install the inverter somewhere else, like inside your garage. In your garage, you can mount it near your power meter. Keep in mind that this can result in your garage heating up when the solar system is running hard.

Any inefficiency is wasted energy, which gets turned into heat. Inverters can get very hot, particularly on the sunniest summer days when production is maximum. When an inverter gets hot, its output goes down. This is a basic fact of electronics in general. Therefore, you want to mount your inverter in the coolest spot possible, meaning out of direct sunlight as much as possible. You also want to make sure that the inverter can get fresh air for ventilation purposes. If you mount an inverter in a tiny closet, with no vents, the air inside will heat up and so will your inverter; it's a cascade effect.

## Tracking mounts

Tracking mounts mechanically move the PV panels over the course of a day so that they directly face the sun at all times. Dual axis trackers change both azimuth and elevation, while single axis trackers only match the azimuth.

Most trackers don't use motors or gears. Very simply, a gaseous refrigerant is sealed within tubes. Sunlight heats the refrigerant on one side, causing it to boil, expand as a gas, and then condense on the cooler side. The movement of the refrigerant causes the panels to move. When the tracker faces the sun, both sides are evenly heated.

Despite the simplicity of tracking systems, they still move (any system that uses moving parts is less reliable), and their lifetimes are unpredictable. Pay attention to the warranty. Sometimes tracking systems get confused. If the day is partly cloudy, the trackers have a hard time, and you may find your panels aiming where they shouldn't be aiming.

Manual trackers are another option. With these, you make the changes with a wrench and some elbow grease. They're sometimes practical for elevation adjustments, particularly with ground mount systems. Several times a year, you go out to your array and physically adjust the tilt angle up or down (lower in the winter, higher in the summer) to achieve a higher energy output over the course of the year. There aren't really any moving parts, other than nuts and bolts, so the reliability doesn't suffer like it does with automatic trackers.

A tracker can increase a system's output by 25 to 30 percent per year, but it also increases the cost of a system, so there's a wash. A better way to increase system output is to simply use more stationary mounted panels.

## Disconnect switches

Disconnect switches are of critical importance, and they need to be mounted within easy reach. Every member of your family should know exactly how to turn the PV system off for safety reasons. If any abnormal behavior occurs in your home's electrical system, shut off the solar system first.

The trend with new inverters is to include the disconnect switches. This results in lower system cost and higher reliability. For more on inverters, see the earlier section "DC-to-AC inverters."

## Wiring and fuse box connections

Wiring, conduit, and connections to your household main fuse box are minor hardware expenses, but they comprise a big chunk of the labor when you're installing a PV system. You want the installers to hide the conduits if they can (in your attic space, for example). PV systems generally use heavy gauge wire (if the wire gauge is too thin, you'll lose productivity, especially when it's hot out), and long wire runs cost more. Remember, all wires are resistors, which means they consume power. The hotter the wire, the more resistance. You need to design your system for worst case scenarios, which means for the hottest days of the year.

The most important consideration is the wire size, and you need to consult tables in the NEC (National Electrical Code) to determine which sizes work optimally for your application. When wire runs are long, larger conductors must be used. The price goes up fast when wires are thicker, so you'll want to try to minimize the runs between the various components in your system.

*Conduit* is a metal or plastic tube or pipe that contains wires and offers protective enclosure. Bare wires and cables are forbidden under most circumstances, so you'll need to use plenty of conduit in your installation. Junction boxes are also used to contain wires, particularly where there are *splices,* or connections of individual wires. A typical PV system entails all the wiring to be contained within conduits and junction boxes.

PV systems must be wired according to NEC (National Electrical Code) standards. This ensures safety and reliability, and if you're going to DIY you need to be well versed in the regulations. In particular, NEC Article 690 covers PV system installations.

Wire connections are the main cause of failures, for a variety of reasons. Here is where a good contractor excels. A single poor connection can result in a very poor system performance. Dubious contractors skimp on wire size because it saves them money, and most homeowners have no idea how thick the wiring should be so they won't know if they're being scammed.

## Utility power meters

New power meters are usually provided by the utility company when the serviceperson comes to your house to inspect your system and connect it to the grid. Conventional power meters are capable of spinning backward, but the utility commonly changes to a special digital meter when you connect to the grid because most solar customers go to the TOU (time-of-use) rate structure, which requires more intelligent processing than a mechanical device is capable of. And even if you stay on the same rate schedule, most power meters are not designed to run backward, so they have to be replaced by new units. The utility company may charge you for the new meter and the installation, as well as a hookup fee. *C'est la vie.* Find out if this is included in the cost of your PV system; some contractors won't tell you that you'll need to pony up the extra bucks, once they're finished with their job, and this can cause undue animosity.

## Installing a PV System

In this section, I describe the process that you should follow when you have a solar PV system installed, whether you do it yourself or hire a contractor. If you opt to do it yourself, the nuts and bolts won't change but expect the process to take more time.

# *All the things you need to do: The process in a nutshell*

You don't want to hurry your decision-making, so expect the entire process of installing a full-scale PV system to take 90 days or more. Too much money is at stake, and you want to make sure that you're doing everything just right. Delays are also inherent in getting building permits, rebate approvals, and so on, and you can't do anything about these.

The following list outlines all the things you need to do:

1. **Perform an energy audit and take conservation steps (see Chapters 2 and 3).**

   Before you even get started on your solar system, you need to figure out exactly how much energy you're using and where it's going. This is particularly important when you're signing up for a time-of-use (TOU) rate structure because, just by doing the audit, you'll be able to glean insights on how to use power differently at different times of the day.

   Some states require an energy audit before you can buy a solar system or before you can collect any available rebates. Why? It's not really in your contractor's interest to help you reduce your power consumption, which would mean you need a smaller (usually less-expensive) system, even though you get a much better payback by finding ways to conserve before you buy a big PV system. Plus it's in society's best interest to use as little power as possible. Hey, the government is subsidizing you, so they have a right to tell you what to do — and besides, they're right.

2. **Determine the utility rate structure you'll qualify for *after the equipment is in place.***

   Your current structure may not be applicable anymore. Can you change structures later if you don't like the one you're in? Can your utility company change structures on you later? You may find yourself in a real bind if you install a system under a certain rate structure but then get a nice letter from your utility informing you that they're changing your status.

3. **Review the physical installation options.**

   How much roof space will a system take up? Do you have a suitable roof, facing approximately south? If not, you may have to ground mount, which is more expensive, plus visually questionable for the neighbors. What condition is your roof in? If you need a new roof, you should probably take care of that first because the roof job will be a lot more expensive if you have to have the PV panels removed (the roofers will certainly not do it) by a solar contractor and then replaced at the end of the job.

**4. Decide how much to invest and how to finance it.**

During the course of your energy audit, you collect a lot of financial information regarding energy costs and how they accrue in your household. You must also collect cost and performance estimates for PV systems, including PV system costs, lifetimes, expansion potentials, warranty, and so on.

The best bet is to call contractors and have a preliminary conversation about these issues (trust me, they'll know all the rebates, subsidies, and tax credits because it helps them sell systems). Don't be shy about exercising these people — if they don't want to help you and answer difficult questions, find somebody who will.

You won't accrue the benefit of tax breaks until your next tax filing, which can take over a year or more. Are rebates payable to you or the contractor? In California, rebates are paid directly to the contractor, so they're not a part of your cash flow at all. Chapter 20 can fill you in on subsidies and loans.

**5. Locate contractors and go out for formal bids.**

Talk to as many contractors as you can. Get them to come to your house and look at your situation in some detail. They can't give you an accurate quote until they do. If they quote you a price over the phone, it's only an approximation. If they tell you it's not, say goodbye right before you hang up.

**6. Choose the best contractor and write the contract.**

At this point, you'll probably have to write a check for a down payment. For more on choosing a contractor, see Chapter 19.

When you discuss the terms of the agreement, make sure to work out the cash flow for the investment, You'll probably have to make a down payment (typically $1,000) at contract signing. Then you'll have to pay somewhere like 20 percent of the remaining balance at the beginning of installation. As soon as the system is in place, the other portion is due. Make sure to find out when the system will begin producing power. You may have a gap in time between your last payment to the contractor and the utility approving your system, at which point you can turn it on.

It's illegal, in most locales, for a contractor to charge for work that has not yet been finished. You should not have to pay a contractor in advance, which means that progress payments should be well defined and should match the work that has been done (not the work that's going to be done). If you think a progress payment is too high, discuss it with your contractor, who will often back down. At the very least, make sure that the work you're being charged for has, in fact, been done. If the contractor is charging you for a system design, ask to review that design.

7. **Wait for equipment to arrive (it's rarely stock), approvals for building permits, subsidies, tax breaks and so on.**

   Expect this to take up to six weeks or more.

8. **Allow for installation and inspections by the county and utility company.**

   Installations typically take a couple of days (ground mounts take a week or more). The county inspectors will look at your system and certify it.

9. **Wait for the utility to put in a new meter and connect to the grid.**

   When everything is ready, the utility company installs a new power meter and officially hooks you up. Now you're in the power generating business. Woohoo!

10. **Get a tutorial on how to operate your system.**

    Your contractor needs to walk you through the entire system and explain the hazards and proper operation. You should be aware of potential problems and how to identify them. At this point, you can watch the display on your inverter cranking out numbers. Your contractor should explain exactly what the numbers mean.

11. **Submit any paperwork to utilities, states, and so on for final rebate payments.**

    Rebates aren't payable until the system is in place and working properly. If your contractor is receiving the rebate directly, you don't need to do anything. If you're receiving it, you want to get it as fast as you can.

12. **Change your household habits to optimize system payback.**

    Up to this point, everything has been educated speculation. Now reality sets in. Is your system operating the way it should? The way it's capable? Are you using power at the right times of the day? Are your savings what you thought they should be? If you're on a tiered rate structure, or a TOU rate structure, you probably need to change some of your consumption habits in order to capitalize. Talk to your contractor about the things you can do, and if the system is not producing the way it was projected, why that may be the case.

13. **Maintain and repair the system.**

    In Murphy's immortal words, "Whatever can go wrong, will go wrong." Unlike most other financial investments, PV system problems are entirely yours to solve. Even if you're under warranty, you have to call the contractor and notify him; he has no idea of knowing when your system is broken.

# Contractor or DIY? Things to think about

You can install your own PV system in many locales. I have seen a number of self-installed systems and they work just fine, and the owners are happy with the results. On the other hand, there are a lot of reasons to use a contractor, namely because of all the complexities involved in a PV system, some of them very subtle. Experience counts for a lot.

Before you go DIY, first consider all the items on this list:

- ✔ You may not be able to connect to the grid with a self-installed system. Before you do anything else, check with your utility company, and your county building department.

- ✔ You may not be able to get insurance for your home with a self-installed system. After all, it entails more risk, and if a system malfunctions it can cause all kinds of problems, the worst being physical injury or a fire. Check with your insurance company.

- ✔ Getting permits and inspections will be your responsibility. Visit your county building department and ask about the range of requirements you'll be expected to meet. It's not a short list, most of the time.

- ✔ Some equipment manufacturers will simply not sell equipment to anybody but licensed contractors. This is understandable; their reputation is on the line and if a DIY installs poorly, the manufacturer's reputation suffers.

- ✔ When selecting equipment, it's imperative that you understand the installation manuals. If they're poorly written, or confusing, don't use that piece of equipment. You can get most manuals over the Internet, even before you buy a piece of equipment.

- ✔ You may not be able to save much by installing yourself. Contractors buy large quantities of materials, and they get much better prices, which they can pass on to you. In addition, installations can require expensive tools that you probably don't have in your garage right now. You'll have to add the cost of these specialty tools to the cost of the job.

You can probably find a contractor who will work with you on an installation. The contractor will design the system and provide the parts and let you physically mount the equipment into place and wire it up. Then the contractor will check out your work. At the very least, I would recommend having a pro do the design, if nothing else.

- ✔ You won't get a ten-year installation warranty if you go it alone. If you install a rack mount system on your roof and it leaks, for example, you're on your own. If you have problems that are strange and difficult, you may have to hire a solar contractor to come in and straighten things out.

✔ If you don't understand electricity, forget installing the system yourself. There are a lot of dangers with a PV system, and you need to understand exactly when and where these dangers can rear their ugly heads.

If you decide to install the system yourself, my overwhelming recommendation is to use a kit, which is the subject of the next section.

## Working with kits

Some manufacturers offer complete kits for solar PV systems. If you're going to install yourself, there are some big advantages in using kits.

✔ The design has been worked out, and it's going to operate the way it's supposed to (given that you mount and connect the equipment properly). PV system design is very complicated, particularly optimizing PV panel strings that feed the inverter. With a kit, the optimization has been done for you.

✔ You'll get all the parts that you need and they'll work well together. When you're finished with your masterpiece and there are parts left over, you'll know you didn't do something that you should have done.

✔ When you design and install your own system, and it's not working up to par when you finish, the manufacturers of the individual components are often reluctant to honor warranties. When you use a kit, the warranty terms are spelled out precisely; basically the only requirement is that you install the system properly, which is easy enough to prove by a simple visit to the site.

✔ With most kits, a customer service number is provided so that when you run into problems an expert can help you through the mess. Terms vary, so understand what kind of support you'll get and whether you have to pay. Typically, there is a time limit to how much they'll provide for free, after which point you have to pay an hourly rate.

✔ The price of a kit is very attractive, compared to buying separate components from separate manufacturers.

✔ Assembly instructions are well written and concise (they better be, or they won't sell many kits). They take you through every step of the process and they spell out the dangers and potential problems.

Of course, you still have to apply for your own permits and deal with inspectors on your own. Even if you're really good with electricity and tools and are capable of installing a kit system like a pro, that doesn't mean you're an expert at working through the permit process, which can sometimes take every bit as much time as installing the system itself. And it can be very frustrating, as most government bodies seem to be set up to serve the government bodies, not the "customers."

# Safety

Above all else, you want your system to be safe. PV systems generate all kinds of dangerous voltages, not to mention the physical dangers inherent in climbing around on roofs and using power tools if you do the installation yourself.

PV panels are dangerous even when they're not connected to anything. The two wires from a PV panel are "hot" whenever sunlight shines on the panels. If you wire a number of panels up in series, the voltages can exceed hundreds of volts, which can literally kill you. If you have turned your PV system off, that does *not* mean it is safe. Any sunlight striking the panels can generate enough oomph to cause major shocks. *I repeat: If utility power is shut off, and you have turned the PV switches off, your system is still dangerous!* Do not open the switch junction boxes, nor the inverter. Let a pro do the work; they know what's safe and what isn't.

The best bet, from a safety standpoint, is to understand exactly what you're doing at all times. With many projects, you can learn as you go. Not so with PV; understand the entire process well before you go.

Following are some general guidelines for ensuring a safe installation:

✔ Be alert at all times. The fact is, most accidents occur when people are daydreaming. Drink your coffee before you begin work, and don't be talking on your cell-phone while you're messing around with your system.

✔ If possible, never work alone.

✔ Use good quality tools, in proper working order. Cheap tools are not cheap and wear the appropriate safety gear:

  • A safety helmet when you climb around on a roof.

  • Eye protection at all times

  • Dry leather gloves whenever possible (sometimes you have to take the gloves off because you just can't achieve enough manual dexterity with gloves on).

✔ Always assume that switches are turned on, and never assume that a switch is working properly. Sometimes switches in the off position aren't off. Never assume your system is safe, even if it's supposed to be.

✔ Understand the principles of grounding. The purpose of grounding is to prevent unwanted currents flowing through people or equipment that they're not supposed to flow through. If you're not properly grounded, you can be killed if something goes wrong.

# Chapter 17

# Is a Full-Scale PV System Worth the Investment?

*In This Chapter*

▶ Understanding the financial viability of a full-scale system

▶ Identify how large a system you need

▶ Figuring out whether a full-scale system is right for you

*W*hen you think about payback for any solar system, you consider a variety of factors, such as how much power the system produces, what your monthly saving will be, and how much the system will cost. This chapter helps you go through specific calculations related to these factors for a full-scale PV system. Of course, you also need to consider other things, like how much your home's value will appreciate and what kind of dollar value you'd put on reducing pollution. For information on these payback considerations, head to Chapter 6.

## Why an Intertie System Makes Financial Sense

In most cases, the financial viability of a PV investment hinges on whether you can install an intertie system. Here's the basic idea: You install a PV system at your house, which connects directly to your household electrical system. With an intertie system, when your PV panels are generating more electrical power than your house is using, your meter spins backward because that excess power is going right back into the utility grid and being used by somebody else connected to the grid. When your solar system is not producing as much power as you're using in your home, the grid provides the difference. It's like a bank account; you can withdraw or make deposits at will. You don't have to worry about how much power you're using at any point in the day. If your solar PV goes down, for whatever reason, you'll have backup.

The following sections explain why a full-scale PV system makes a lot of financial sense in general. The remaining sections of this chapter take you step by step through determining whether such a system is a good idea for your particular situation.

## Exploiting every bit of energy your system generates

Here's the simple reason why intertie is so important from a financial angle: You exploit every ounce of the energy that your system generates. To understand this point, consider a solar hot water heater (refer to Chapter 12). When you go on vacation for two weeks, the equipment sits idle: It's not paying you back one single dime. With intertie, you benefit from every photon that strikes your system. You get the most bang for the buck for that expensive equipment you're investing in. When you go on vacation, the solar PV system is subsidizing your vacation.

Intertie is such an important concept that state and federal governments are legislating that utility companies must allow their solar-producing customers to tie in to their grids. Utility companies, for their part, don't necessarily want these types of customers; their technical risk increases, and their revenues decrease. Plus solar PV entails more equipment and more complex power meters (conventional meters aren't calibrated to spin backward). When you install a solar PV system, the utility will probably have to change your meter and, in many localities, it can't charge you for the new meter.

Most intertie systems won't work when the grid is down. This concept is counterintuitive: You install your own power production equipment, so you should be able to have power regardless. But the fact is, the utilities don't want you putting power into the grid when the grid is down because it endangers line workers and may actually destroy transformers and other power equipment. The safest and best way to eliminate this risk is to require that PV intertie producers use equipment that won't operate when the grid is down. Although you can purchase PV system that will work during a power outage, these systems there are more expensive and probably not worth the extra cost.

## Understanding net metering

*Net metering* means that you can sell your excess power back to the utility company. It's one thing to feed the power back into the grid, it's another matter being paid for that excess power. The way in which you get paid is very important, and I discuss this in detail in subsequent sections.

In addition, under most net-metering laws, utility companies must pay you the same retail rate they charge their own customers for power at any given time of the day.

Prior to deciding on solar PV for your home, find out how much the utility will pay you for power that is fed back into the grid. If it's less than what they're charging you, you're better off using all your generated power and not feeding any back into the grid. To do this, you have to use more power during the afternoon hours when your solar PV system is producing the most output.

When PV first came onto the market, the energy consumed in the production process for PV panels was greater than the energy that the panel could put out in its lifetime. This contradiction was acceptable for NASA applications but not at all economical for residential consumers. Over time, production costs decreased to the point where solar power became a viable financial investment, but only under the best circumstances. In order to encourage widespread use, the government stepped in and initiated subsidies and tax breaks. But the biggest benefit of all came when net metering was legislated into law.

# Determining System Size

When thinking about system size, the problem generally comes down to one of economics; which system size provides the best return on investment? For those intertie customers who want to offset their entire electric bill or for those who are off-grid and have no choice, sizing the system comes down to calculating how much energy you use in your home, and when you use it.

For those who have been in their homes for a few years, the problem of determining usage is straightforward: Get your utility bills for the last few years and make a list of the average monthly usage. If you're planning on adding loads, like a new swimming pool, you need to calculate how much that load will consume and add it to your list. Most appliances come with a label that tells you exactly how much power that appliance draws. A swimming pool pump is around 2 HP (horse power), which is around 1,500 watts, or 1.5 kW. If you run it for 10 hours per day, you'll consume 15 kWh's per day, or around 450 kWh's per month. If you're paying upper tier prices, say 44 cents per kWh, this will cost you $198 (wow). Perhaps you'd be better off running through a sprinkler.

If you're building a home, you need to anticipate how much power you'll be using. Draw up a list of the loads and how much you expect to be using them. This task isn't difficult, just time consuming. You can get a list of power loads for household equipment in my book *Energy Efficient Homes For Dummies* (Wiley).

Once you determine your loads or how much energy you use in your home, the next step is to determine PV system size. Do you want to offset your entire electric bill or just a portion? Are you most concerned with pollution emission offset, or return on investment? The biggest problem you will be faced with is predicting future energy prices. Historically they have risen around 7 percent per year, but this is probably going to increase, especially if cap and trade policies are instituted. When you do the financial analysis in the following sections, be sure to estimate this number. Most contractors will provide you with a financial analysis, and you can tell them to use a particular inflation rate. The best bet is to do several financial analyses, using different inflation rates; that way you can see how much future energy price increases affect the viability of a solar PV investment.

# Predicting System Production

How many kWh's of electrical energy is a solar PV system going to provide you, over the course of a year? The system's output will vary over the course of a day, and from month to month; that's unavoidable. Ideally, you want to predict how much power you'll get each month.

The local climate determines how much sunlight you can expect over the course of a year. A PV system in Seattle, Washington, where it's cloudy and rainy, will be harder to justify than a system in Phoenix, Arizona. You can account for your local climate's lack of sunshine by getting a larger system, but doing so obviously costs more, and the payback is not so good. (For more on estimating solar potential, see Chapter 5.)

These factors also affect production: Where on your roof will you mount the solar panels? How will you mount them? Which direction will they face? Are there shade issues that you can deal with, or are the shade issues a fact of life that you must live with?.When you install a solar system, your goal is to maximize production.

A number of widely used computer programs are available that you can easily work yourself. The most common is PV Watts (www.pvwatts.org). To get a projection of your system's performance, enter the data requested, outlined in the following sections. To give an idea of what you can reasonably expect from solar PV, I've included a typical PV system at a typical size in these sections. (I get into a lot more detail about hardware in the next chapter, but the system used as an example here should give you a ballpark idea of what you can expect from state-of-the-art equipment). In the later section "Calculating the Cost of a System," I provide examples of how to calculate payback, and I use this system as a basis.

Due to the complexity of rate schedules, it's usually not easy to determine how much your solar PV system will produce in terms of dollars saved. Contractors have computer programs that can generate detailed analyses of your payback, but there are a lot of assumptions built into these programs and you need to understand the assumptions that are being made if you want to gauge the accuracy of the claims. Contractors want to make your solar PV investment look as good as possible, for obvious reasons, and you need to be able to judge whether you'll get as much out of your system as they claim you will.

# Common terminology

You'll need to understand some basic industry terminology in order to work with solar PV systems.

✓ **Load:** Any device that uses electricity. An inverter is a load, as is a washing machine or a light bulb. When determining system size, you add up all the various loads, and how much they are being used in your home.

✓ **Volt:** The unit of force, or electrical pressure, that causes electrons to flow through a conductor (wires and electronic equipment). Volts are abbreviated as V, or sometimes E (which stands for electromotive force). Most household electrical systems operate on 240 VAC, or 120 VAC.

✓ **AC:** Alternating current. The flow of electricity goes in both directions, back and forth, and in household electrical systems it flows back and forth 60 times per second (referred to as Hertz, or Hz). In some regions, the line voltage and frequency (the number of Hz) from the grid may vary quite a bit, and this can cause problems with some of your household loads. It may also require you to specify an inverter that can handle the fluctuations.

✓ **DC:** Direct current, where the flow of electricity is in only one direction. The output of a solar panel is DC. Since your home uses AC, this DC current must be converted into AC to be useable, and this is what an inverter accomplishes.

✓ **Ampere, or Amp:** The amount of current that is flowing, due to the presence of a voltage. Amps are denoted by I (for intensity of current).

✓ **Resistance** (denoted by R): The amount of impedance to electrical flow. The higher the resistance, the lower the current for a given voltage.

✓ **Watt:** A unit or power, and is equal to Voltage times Current, or I X V. It's an instantaneous measurement; power can vary from second to second, as you switch the various loads in your home off and on. The output of a solar PV system is listed in watts, and this varies with the sunlight striking the panels.

✓ **Watt-hour:** A unit of energy, and this is what your solar system produces over a period of time. At any given time, your system is outputting a certain power level (watts), but over time it produces energy, and ultimately energy production is what you are after. Typically, we're more concerned with kWh, which simply means one thousand watt-hours. Your utility bill charges you for how many kWhs you use per month (Chapter 17 explains the various rate structures, or the way your power company charges you for the energy you use). A 100-watt light bulb burning for one hour uses 100 watt-hours of energy, or 0.1 kWhs.

## Panel types

As Chapter 16 explains, there are a variety of panel types you can choose from: traditional, flexible, and so on. Each type has its positives and negatives, and any can be the best choice for any particular scenario and payback goals.

*Example system:* Panels: Quantity 24 Sharp 198 U1F panels, at 198 watts DC output. These are very popular panels, and Sharp is one of the world's leading suppliers. The technology is sound and proven, and Sharp's track record is impeccable.

## Inverter type

Inverters vary in terms of efficiency and operating parameters (see Chapter 16). You'll find, using PV Watts, that given the same panel array, different inverters output different amounts of energy over the course of a year. You want to find an inverter that maximizes energy output, given your particular parameters.

*Example system:* Inverter: Quantity 1 SMA America 4,000-watt inverter. SMA is one of the leading suppliers of inverters. They're German made, and Germany has been on the leading edge of solar power for many years, so the technology is proven and sound. Efficiency is high, and reliability is excellent.

## Tilt angle

The tilt angle of your system will be optimal at your latitude. For example, if you're at 38 degrees north, that tilt angle gives you the most production over the course of a year. For most PV customers, however, the tilt angle they're stuck with is given by their roof pitch. You can use special mounting racks to change the tilt angle, but they're costly and look ugly. Fortunately, the output of your system is relatively insensitive to tilt angle. For a tilt angle that's 12 degrees off of optimum, system output goes down only a few percent. Most roof pitches support a good PV output, regardless of their pitch angle.

*Example system:* The panels are mounted at a tilt angle of 28 degrees, which matches a typical roof.

## Azimuth

The azimuth angle is more influential on system output than tilt. True south is the best azimuth, but 10 degrees more or less won't affect your production more than a few percent. Southwest is better than southeast. If your roof is facing southwest (225 degrees), expect a decrease in production of around 6 percent. If your roof is facing due west, expect a decrease of 16 percent (some government subsidies won't pay if you're facing due west or due east because the system won't be outputting enough energy). If your roof is facing southeast or due east, expect more of a decrease than if it's facing southwest or west. Bottom line, you're better off, production wise if the system cants to the west as opposed to the east.

The government pays you, in the form of subsidies and rebates, to install solar, for the obvious reasons. If your system doesn't produce enough, the government may not deem it worthy of these financial incentives. California, for example, doesn't give a rebate if the system faces due west simply because such a system won't put out enough energy to merit a rebate; the economics are no good.

*Example system:* The panels are mounted facing due south.

## Shade levels

Shade, which can come from not only trees and power wires but chimneys and vent pipes and neighbor's homes, is by far the worst enemy of solar PV. Fifty percent shading on an array can result in 80 percent of the power output being interrupted. While shade can often be engineered around, the cost rises. For instance, if you have a big vent pipe in the middle of your roof, the solar array may need to be divided into two separate arrays which both avoid the vent pipe.

Be aware that shading varies over the course of a year. The sun is lowest in the sky in the winter months, and so there will likely be more shade in the winter than the summer. The production programs ask for shading in each month, and you'll have to estimate that as best you can. Be conservative in your estimates.

*Example system:* There is minimal shading, which means less than 5 percent, all year long.

## Distance from the roof

In the calculation programs, you need to specify the distance from the roof — longer distances give you more power output, but they're uglier. The most common distance is three inches.

*Example system:* The panels are mounted three inches from the roof.

## Rated DC and AC output

The number and type of panels determine the DC output of your system. This power is fed into the inverter, which turns it into AC. The production programs tell you what the AC output of your system is (the AC output depends on the system's efficiency).

*Example system:* The rated DC output is 4.752 kWs, and the AC output (that which comes from the inverter and is fed into the power grid) is 3.908 kWs. The difference is due to inefficiencies and conversion losses (the difference may seem big, but it's actually very good).

## Expected energy output

After you enter in the data for your system, the program performs its calculations and gives you the results. In the example system, the expected output energy per year is 6,845 kWhs, of which 60 percent is generated in the months of May through October (summer months, in utility vernacular). Here's how the system will output this 6,845 watts by month:

| *Month* | *Output (Percentages)* |
| --- | --- |
| January | 4.5 |
| February | 6.2 |
| March | 8.5 |
| April | 9.7 |
| May | 10 |
| June | 10.4 |
| July | 10.7 |
| August | 10.9 |
| September | 10.2 |
| October | 8.9 |
| November | 5.7 |
| December | 4.3 |

*Note:* This breakdown will change depending on a number of factors, such as month to month shading, tilt angle, azimuth, and so on. But for the vast majority of cases, this type of breakdown is typical.

The numbers represents the best that may be achieved with 24 panels outputting 198 watts per panel. You can scale these numbers up or down and get a good estimate of what to expect from the size system you're planning on installing.

I'm going out on a limb here, but you can expect a system like the one outlined in the example to cost between $6.50 and $8 per DC watt. The total cost depends on the installation, of course. To determine the net price (how much of the cost you actually pay), subtract all the subsidies and rebates from the total cost. A typical net price may be around $4 per DC watt.

# Calculating Monthly Savings

Once you know how much your system will produce, the next problem is to translate this into monthly utility bill savings. To perform this calculation, you need to understand precisely how your utility charges you for power, which means you need to know your rate structures.

In many localities, you have some choice about which type of rate schedule you can sign up for once your solar system is in place. The structure you choose will impact your payback, so it pays to understand the differences between the various schedules.

Most solar PV contractors use complex programs to generate not only a prediction of your solar PV system's output energy (as explained in the preceding section), but also a financial analysis of how much you can expect to save with the system. You can generate these calculations on your own, or use PV Watts (enter "PV Watts" into your internet search engine). These programs rely on an assumption of power rate increases. Try some different rate increases and you'll see that your monthly savings will vary quite a bit, depending on the level of power cost inflation.

## Types of rate structures

In order to understand exactly how your solar system will profit you, you need to at least understand the rate structure you'll be operating under after your system is finished. In order to calculate how much you can save by installing a solar system, you also need to understand the rate structure you're working with now, so that you'll be able to compare the before and after results. Here are the most common types of rate structures:

✔ **Simple:** Simple rate structures charge $x$ dollars per kilowatt-hour. No problem. The first kilowatt-hour you buy is the same price as the last. Unfortunately, this structure is rare, and the trend is to abandon it because it doesn't encourage conservation.

✔ **Seasonal:** Seasonal rate structures change rates in summer and winter. Summer rates are almost always higher than winter. You may not even know it, but you're probably on a seasonally adjusted structure right now.

✔ **Tiered:** In a tiered rate structures you get a baseline energy usage, say 600 kilowatt-hours per month. If you use more than this, the rate goes up, sometimes punitively. Here's a sample tiered rate structure:

| Percent of Baseline | Monthly Energy Use | Cost Per kWh |
|---|---|---|
| 0–100% | 0–600 kWh | $0.11531 |
| 101–130% | 600–780 kWh | $0.13109 |
| 131–200% | 780–1,200 kWh | $0.25974 |
| 201–300% | 1,200–1,800 kWh | $0.37866 |
| Over 300% | 1,800 kWh | $0.44098 |

Most tiered rate structures also have different rates for summer and winter. Compare your February bill with your August bill, and you'll probably find that your power costs more in August. Summer rates are sometimes a lot more. The summer season is typically May through October, six months.

✔ **Time-of-use (TOU):** TOU rate structures have different rates depending on what time of day you use your power. Here's a typical example:

| Time | Summer Rates | Winter Rates |
|---|---|---|
| Peak (noon to 6:00 p.m.) | $0.29372/kWh (ouch!) | $0.11472/kWh |
| Off-peak | $0.08664/kWh | $0.08966 per kWh |

This rate structure requires a special power meter with a clock and two different totals: peak usage and off-peak usage. Obviously, using power in the afternoon, in the summer, is very expensive. But here's the key; your solar system will be outputting its maximum amount of energy during summer peak hours. If you can manage to use very little power in your home during summer peak times, your power meter will be running backward at the summer peak rate (in other words, it'll be running backward very fast). Figure 17-1 compares time-of-use to energy production for a typical PV system.

✔ **Combined:** Combined rate structures can get very complex. In Northern California, intertie PV customers may choose to go to a tiered, TOU, seasonal rate structure as in Table 17-1. Note, in particular, the peak-to-off-peak ratio for summer use is over three times.

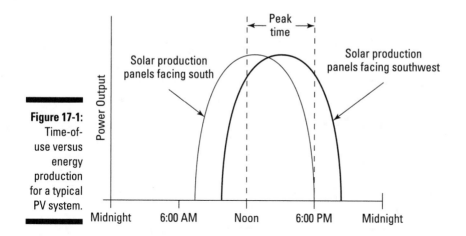

**Figure 17-1:**
Time-of-use versus energy production for a typical PV system.

| Table 17-1 | Rate Structures in Northern California | | | | |
|---|---|---|---|---|---|
| Season/Time of Use | Tier 1 | Tier 2 | Tier 3 | Tier 4 | Tier 5 |
| Summer, peak hours | $0.29372 | $0.29372 | $0.39105 | $0.48102 | $0.52817 |
| Summer, off-peak hours | $0.08664 | $0.08664 | $0.18397 | $0.27394 | $0.32109 |
| Winter, peak hours | $0.11472 | $0.11472 | $0.21205 | $0.30202 | $0.34917 |
| Winter, off-peak hours | $0.08966 | $0.08966 | $0.1869 | $0.27696 | $0.32411 |

Figure 17-2 shows this data in visual form.

When you install a PV system, you may end up paying your utility once a year instead of monthly. You'll get a summary statement each month, with a small service charge (always those service charges, or as they like to call them "surcharges"). Here's the reasoning: most utilities will not pay you if you produce more power than you use. But there will inevitably be individual months where you use less than you produce, so the way to deal with this is to use an annual billing cycle. This is either advantageous, or a pain in the butt, depending on how your budgeting works. You won't have any power bill to speak of for 11 months, then you'll get a big fat one, if you're using more energy than your system is producing.

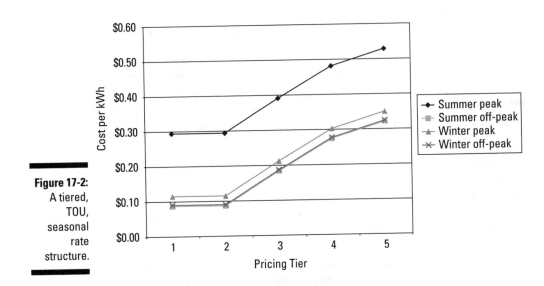

**Figure 17-2:**
A tiered, TOU, seasonal rate structure.

The future outlook of energy prices may be the single biggest factor in how much you choose to invest in a PV system. Whatever your rates are right now, you can safely bet they'll rise. Historical data suggests around 7 percent per year (an average over the entire country). But they rise more in some areas than others, and with the political winds blowing the way they are now, huge rate increases may be on the horizon. It's also important to understand how they rise. For instance, if you're in a tiered rate structure, you can expect that the upper tiers will rise faster than the lower tiers. In early 2008, PG&E (Northern California) increased their highest tiers by 23 percent, while the lowest tiers actually went down less than 1 percent. Since you'll be eliminating the upper tiers with your solar system, this implies that a solar investment is far better than intuition would suggest.

## Analyzing monthly savings in detail

The best way to illustrate how to do payback calculations is to jump right in and look at some examples. You'll want to crunch numbers for an entire year, but I present one month calculations just to keep things as simple as possible. You'll be able to repeat the same process for each month of the year and come up with annual totals.

I use the system that I presented in the earlier section "Predicting System Production," namely quantity twenty-four 198-watt panels and one 4,000-watt inverter with an annual output of 6,845 kWhs.

To perform these calculations on your own, you need to get copies of your power bills, preferably for the last three years and then average the results. Call your utility company, which can provide you with the data you need (the customer service number is listed on the bill).

## Flat rates

Calculating monthly savings for flat rates is simple. If your power rate is 0.11 cents per kWh, you'll save $753 per year with the sample system (6,845 kWhs times 11 cents). If you want more detail, you can use the monthly breakdown of power output presented in the earlier section "Types of rate structures" to determine the savings you can expect each month. Then to calculate how much per month you'll save with your system, compare the monthly cost savings with your historical record of monthly utility bills.

Say, for example, that you pay a net price of $4 per DC watt, or a total net price of $19,000 for your system. Your payback will be 25 years, excluding utility cost inflation. With a 7 percent inflation rate, your payback will be more like 16 years.

## Tiered rates

For tiered rate, do a breakdown of the month of July, where a home uses 2,200 kWhs of energy. Using the tiered rate structure defined previously, the monthly power bill will be:

600 kWh @ $0.11531 = $69

180 kWh @ $0.13109 = $23.60

420 kWh @ $0.25974 = $109

600 kWh @ $0.37866 = $227

400 kWh @ $0.44098 = $180

The total bill is $608.60.

Our model solar system outputs 10.7 percent of 6,845 kWhs in July, or around 732 kWhs, which saves all 400 kWhs in the highest tier (@ $0.44098), plus another 332 kWhs in the second highest tier (@ $0.37866), for a total monthly savings of $305.70 (The net bill will be $303). Note that the solar system only produced one third of the monthly usage, yet saved half of the power bill. This illustrates why tiered rate structures work best with solar PV systems.

You can do a simple calculation like this for each month of the year, and come up with your annual expected savings. You'll save less in the winter, when your solar system is producing less. If you have a gas heater in your home, your electric bill will be lower in the winter and you won't be paying upper tier prices, so the winter savings will be even lower.

To illustrate my point that a smaller solar PV system represents a higher return on investment in a tiered rate structure, note that a solar PV system with twice the capacity as the one I've used here (at around 1.9 times the price), will output twice the kWhs during July, or around 1,464 kWhs. Performing the same calculation as above yields a monthly savings of only $521. This is only 70 percent more than the smaller system, while the equipment costs almost twice as much.

With the standard sized system, the payback is now on the order of 8 years, with energy inflation rates of around 7 percent. This is a great investment.

### TOU

A customer is on the tiered rate structure, and they decide to switch to a TOU, tiered rate structure when they install solar. Now we're really going to get a good payback.

The system will generate 732 kWhs in July, same as before. Assume that the customer can use 80 percent of their power in off-peak times (this is reasonable, given an aggressive commitment to conservation and efficiency during peak hours). So there will be only 20 percent of 2,200 kWhs, or 440 kWhs used during peak, and the rest in off-peak (1,760 kWhs). The solar system will generate around 75 percent of its output during peak hours, for a total of 549 kWhs. It will also generate 183 kWhs in off-peak. Net consumption during peak hours is now 440 kWhs minus 549 kWhs, or a net backward meter spin of 109 kWhs (this is typical; during peak hours in the summer, the net meter spin is backward, meaning the power company owes you money). In off-peak times, the net usage is 1,760 kWhs minus 183 kWhs, or 1,577 kWhs.

The power bill will now be:

Peak:

-109 kWh @ $0.29372 = -$32

Off peak:

600 kWh @ $0.08664 = $52

180 kWh @ $0.08664 = $15.6

420 kWh @ $0.18397 = $77

377 kWh @ $0.27394 = $103

For a total of $247.

Add the two yields for a grand total of $215.

Compare this to the net bill with the tiered rate structure of $303, and you can see how TOU saves a lot more when you go solar. All you have to do is manage your power consumption during peak hours.

Don't expect to achieve this kind of savings all year round. But note that the winter rates in TOU are lower than they are in the tiered rate structure, so one's overall net bill be lower all year round. The biggest savings occur in the summer months when the meter spins backwards during peak times at the high rate.

This represents a payback period of less than six years, which is an extremely good investment.

Note also that doubling the system size is wiser with TOU than it is with a tiered rate system because the return on investment doesn't decrease nearly as much. When you double the system size in a TOU structure, you'll be making your meter spin backward during peak hours quite a bit more, and this pays back better. In fact, by doubling the system size, the entire power bill can be erased. This was not the case with the tiered rate structure.

If you orient your panels southwest, you'll maximize the amount of solar production you'll receive during peak hours. If your panels are facing southeast, or east, you won't get much production during peak hours.

Even if you don't go with a solar PV system, you may find that you can save on your monthly power bills simply by going with the TOU rate schedule. You can do the same calculations outlined previously and compare your current rate schedule results to those of the TOU. You will need to make an assumption about how much power you use during peak and off-peak times. Realistically, don't use a number more than 90 percent, because you'll inevitably be drawing power during peak (refrigerators, phantom loads, negligent children, etc.) Here's a common scenario; you buy a copy of my book *Energy Efficient Homes for Dummies*, and save an immediate 15 percent on your power bills simply because you are now an energy conscience conservator. Change to the TOU, use very little power during peak, and you could see your power bills decrease by over 25 percent, with no investment other than a few hundred dollars for energy efficiency equipment.

### Other rate structures

My goal so far in this section has been to give you the tools you need to calculate your expected cost savings with the various rate schedules. There are, in fact, a multitude of different rate schedules that take into account medical conditions, agricultural situations, and the most common rate divergence is when a house is powered by both gas and electric. But they're all predicated on a combination of the schedules I have defined previously. The dynamics are the same, while the particulars may vary. I have given you the tools to calculate your expected cost savings with any rate schedule that you may be faced with.

You need to find the rate schedules that you operating under now, and the one that you will be using when you install your solar PV system, and compare the bills before and after. It's as simple as that.

# Calculating the Cost of a System

How much you pay for your system is of critical importance in determining whether the investment is worth it, of course. You want to get the most output capacity for the least net price. The business is competitive, so finding the lowest price should be straightforward; just get a number of bids from reputable contractors.

Prices fluctuate quite a bit from year to year, and even from month to month. Subsidies and rebates change, and market conditions affect availability of panels, in particular. The price of inverters holds steady, as does the price of the support equipment. Contractors will lower their prices due to heavier competition, and the trend is toward lower profit margins because so many new contractors are entering the game. Expect to pay anywhere between $4 per DC watt, up to $6.50 (this assumed the 30 percent Investment Tax Credit (ITC) is in place, which there are no guarantees). Economies of scale aren't a major factor. If you buy a system with double the size, expect around 190 percent price increase. A major portion of the contractor's added value is in fixed costs, which are county permits, travel to and from the site, and so on.

Some contractors quote prices per DC watt, and some use AC. Because the DC number is larger, it makes the per watt price look smaller. To make matters worse, there are different measurement standards that can result in different per watt prices, even though the quoted systems may be identical in technical terms. Ultimately, what you're interested in is the cost per kWh over the lifetime of the system, which can be very tricky to calculate because of the way rate structures work (see the earlier section "Types of rate structures" for details). If at all possible, get contractors to bid the same exact system, with the same mounting arrangements and support hardware. Then you can compare apples to apples.

If you choose to install the system yourself, you'll save labor costs, but you may have to pay more for the equipment because contractors can buy it in huge quantities and get discounts that they pass on to their customers.

# Chapter 18

# Divorcing the Grid: Going Solo with Solar Power Systems

*In This Chapter*

▶ Looking at the pros and cons of living off-grid

▶ Specifying the right solar power systems

▶ Looking at a real-life scenario

*O*ff-grid means you're not connected to a public utility for your electrical power. The term usually also means you're not connected to utility gas, water, and sewer. However, in this chapter, I address only electrical power because that's the angle that solar can solve. Here you find out what to expect when you go off-grid, how to evaluate your system needs, and what an off-grid system may look like.

## Understanding Off-Grid Ramifications

Here are a few reasons for going off-grid:

- ✔ **No utility power is available, period.** Maybe you're living in a remote location.

- ✔ **Bringing utility power in costs way too much.** If you're off the beaten path, you can probably get the utility company to bring power in, but you have to pay for the long line lengths and poles, trenches, and the like. This job can sometimes cost more than $100,000.

- ✔ **You just want to be off-grid because of the allure of independence.** Perhaps you just plain don't like utility companies and abhor paying their monthly bills. Perhaps you love the idea of going completely green. Maybe you just want to be able to gloat during power blackouts when the rest of the world goes black while your power is still on.

You have a number of choices when you go off-grid. Propane works well for heating and cooking and refrigeration. Wood stoves are popular for remote houses and cabins because there's usually a ready supply of firewood. But at some point, you're going to want to plug in your television (imagine life without a TV!), and the choices are more limited. Gas-powered generators are inexpensive and can put out a lot of power when needed. But they're noisy, cumbersome, and smelly, and they pollute quite a bit.

The majority of off-grid homes use some form of solar: wind turbines, hydro-power, hot water heaters, or PV systems. Off-grid systems include batteries because it's impossible to arrange a lifestyle to use power as it becomes available, so some means of storage is needed. They also include inverters, which transform the voltages output by wind turbines, hydro alternators, and solar PV panels into voltages that are commonly used in household applications. The following sections describe the fundamental system architectures and suggest ways to further pursue off-grid solar power systems.

## Considering higher costs and maintenance

Chapters 16 and 17 detail PV electrical generating systems connected to the grid (*intertie* systems). One of the major reasons for installing such a system is that it lets you sell your excess power back to the utility company. The bottom line is that an intertie system operates at 100 percent of its productive capacity because someone is always using energy as it's being created.

Off-grid solar power systems provide electrical power, but it's relatively expensive. When you install an off-grid solar PV system, you can't get anywhere near 100-percent utilization, plus you need batteries, charge controllers (which control the battery functions), and so on; all this extra equipment is expensive, and it requires a lot of maintenance.

You need to understand what you're doing with an off-grid system; you can't just plug and play. You have to take care of batteries, monitor system performance, and adjust your habits on a daily basis depending on how much power your system is generating. In general, off-grid PV systems cost at least two times as much per kilowatt-hour as an intertie system, and they take ten times more attention on your part.

Most banks won't touch an off-grid home. You may not be able to get a first mortgage, and second mortgages are even more prohibitive. If you ever want to sell your house, off-grid will drive away 99 percent of potential buyers, and the higher cost of a mortgage (if one's even available) will be reflected in a lower selling price. If you have a choice, off-grid is not an investment at all — it's a discretionary, luxury expense.

Standalone systems are generally subject to rebates and subsidies, just like intertie PV systems and hot water systems. See Chapter 20 for details on what you need to know about financing these types of systems.

## Looking green

From a pollution standpoint, you can do the world more good by using an intertie system because you'll be generating more usable power for a system of a given size. Each kilowatt-hour of energy you generate with an intertie system offsets that much energy from the grid — it's a one-to-one relationship. This isn't true with an off-grid system; a lot of the potential energy generation is wasted simply because you don't end up using it. (If you're on vacation, generating capacity is completely wasted; if you don't use as much power as your system generates, you're wasting potential production.) There is a fine point here; this is different than wasting energy. This is wasting the potential to produce energy, and while this entails no pollution, or costs, you are forego-ing potential gain, which changes the economics. The best bet is to share your system with the world, if at all possible.

Carbon dioxide is an important pollution culprit these days, but it's not the only one. Groundwater pollution is a growing problem. Waste management is expensive, and it increases everybody's tax burden. If you aren't really savvy with your batteries, the lifetimes can be atrocious, so the pollution problem is increased not only by the mere use of batteries but also by improper use. I give a brief tutorial on batteries, but you need to read the operating manuals that come with your off-grid system, pay attention to the monitors, and discover how to work optimally with your batteries.

On the flipside, most people who choose to go off-grid are environmentally conscious and competent when it comes to using their resources. Conservation and proper usage come naturally with the off-grid lifestyle.

# Surveying Off-Grid Solar Options

Off-grid PV systems are expensive, so find any means you can to reduce the energy requirements. Most off-grid houses use a wide range of energy resources, in sharp contrast to the typical all-electric suburban home. Solar hot water heaters are always good candidates because they're cheaper per kilowatt-hour than an off-grid PV system, and solar lighting systems are always wise. This variety can be an advantage in that you aren't completely disabled with power blackouts. In fact, you'll be completely impervious to power outages, and even if one of your resources goes down, the majority of your lifestyle will still be intact.

Off-grid solar electrical systems all use the same basic components, with other elements added according to need. Figure 18-1 shows a typical off-grid system.

Here are the functions of each part:

- ✔ **Charge controller:** The charge controller feeds current into the battery bank at the required voltage. Good charge controllers draw the best performance out of the batteries and are very important for economics because they influence efficiency.

- ✔ **Battery bank:** The battery bank is typically made up of six or more individual batteries connected with stout cables in either series or parallel arrangements.

- ✔ **Inverter:** The inverter changes DC to AC voltages suitable for use with household equipment. An inverter is optional if you use DC loads exclusively.

- ✔ **DC loads controller:** You may be using both DC (boat, RV, and auto appliances) and AC loads (standard household appliances). The DC loads controller maintains the proper currents and voltages into the DC loads.

- ✔ **AC generator:** As a backup power supply, the AC generator isn't strictly necessary but is usually part of any off-grid system in order to prevent blackouts when the sun is weak for extended periods of time.

- ✔ **Transfer switch:** The transfer switch alternates the power source between either the inverter output (when battery power is available) or the AC generator.

- ✔ **AC loads controller:** This device includes appropriate fuses and switching means and maintains the voltages and currents used by the AC appliances connected to the system.

The following sections outline some fundamental choices with regards to your PV system composition.

## AC or DC

Which type of current you choose depends on what you want to run. If it's just a few lights at night, with a coffee maker and a fan or two, DC is fine. However, the market for DC appliances is far smaller than 120VAC, so you may go for AC if you're using standard household appliances (which is the most common way to go and is cheaper and better because of the widespread availability of AC appliances compared to DC appliances).

DC, which is more efficient because batteries use direct current, is usually the choice for small cabins and small power systems. You can use DC appliances for RVs and boats, so envision your cabin like a big RV, and you get the picture. But DC also requires larger wire diameters, which can be very costly if you need to run lengths of more than 50 feet or so.

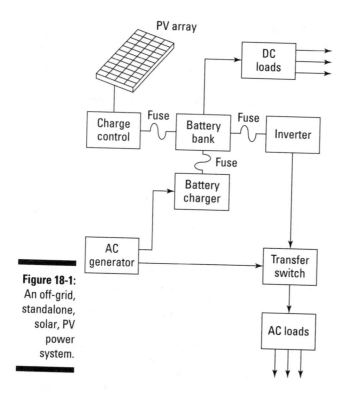

**Figure 18-1:**
An off-grid,
standalone,
solar, PV
power
system.

If you use DC, you need to decide the voltage. 12VDC is the standard used for RVs, boats, and cars, but because it's the lowest voltage, it requires the most expensive wiring (wires need to be heavier gauge to support the higher current draws). Other choices are 24VDC, 48VDC, and so on, but these become progressively more dangerous for electrical shock hazards, and finding compatible appliances is harder. With AC voltages, the universal standard is 120VAC.

# Understanding Batteries (Familiarity Breeds Contempt)

After you install a PV intertie system (see Chapter 16), you can completely ignore it for the most part. Keeping the solar panels clean is about the extent of your maintenance, and you don't really need to do that.

But when you install a system with batteries, you have to stay on top of things. The battery or battery pack is the core of any off-grid system, and it drives the system's cost. All action comes and goes from the battery, and much of the safety and control equipment is designed to protect either the battery or the balance of the system from the battery. You absolutely have to understand batteries, or you'll end up paying an arm and a leg for new ones all the time and you won't get decent performance out of the ones you have.

The following short primer is designed to give you an idea of what you're going to be facing, but it's nowhere near complete. After you buy your system, thoroughly read and understand the detailed directions that come with it.

## Battery types

A wide range of different battery types are available, each with different performance characteristics and varying costs. Battery technology is changing very fast, and new types are constantly coming onto the market. It's impossible to describe the myriad types here. If you need batteries, contact any dealer, such as Real Goods at www.realgoods.com, and ask for technical assistance.

By far, lead-acid batteries offer the best performance versus cost for large energy-storage applications. These are the same type of batteries used in autos, but auto batteries have entirely different internal constructions because they're designed to output a huge amount of current for only a small amount of time; household energy storage applications, on the other hand, require relatively low-power output for long periods. Also, the depth of charge is different for auto batteries (*depth of charge* is the amount of total capacity a battery is designed to lose between charges). Auto batteries yield very little of their total capacity, but household storage batteries are designed to go deep into their total capacity.

## Capacity and efficiency

*Battery capacity* is a measure of how much total energy a battery can hold. Ratings come in amp-hours (Ah). To get the total energy capacity, simply multiply this number by the voltage. Thus, a 100 Ah battery at 12 volts yields 1.2 kilowatt-hours of energy.

In general, the larger the battery capacity, the larger and heavier the battery is. However, the faster you pull the charge out of a battery, the lower the capacity, so the capacity is really useful only for comparing different batteries. It's not a true measure of what you can get out of a battery. (Sound so complex it's almost impossible to get a good handle on things? You're catching on.)

Batteries are also inefficient. When you put a certain amount of energy into them, they lose a percentage. When you take a certain amount of energy back out, they lose another percentage. The average loss is about 20 percent, meaning that for every 100 watt-hours you put in, you can get only 80 watt-hours back out. But inefficiencies increase with age, which means your system capacity also decreases with age.

Also, the higher the voltage, the more efficient the system. Batteries are available in either six or twelve volts, and you can connect them in series or parallel (see Chapter 4) to get 6VDC, 12VDC, 18VDC, 24VDC, and so on. If you're using an AC inverter, with AC appliances, it won't make any difference which battery voltage you decide on, so ask your dealer what he recommends and make sure that the rest of the system is compatible with that voltage.

## Maintenance

Batteries self-discharge at a rate of around 5 percent per month, so when they sit around, they grow useless. If they're dirty, with oily or corrosive tops, they can discharge at a rate of 5 percent per week. Furthermore, battery internal chemicals can crystallize, resulting in a resistance to recharging. The solution is to discharge and charge on a regular basis. If you have an off-grid PV system with batteries and you go on vacation for an extended period of time, your batteries will not appreciate it at all. They'll miss you so much that they self-destruct.

Batteries are like your body: They need activity to stay in peak health. Too much activity, and they get worn out. Too-intense activity, and they can be damaged. The best bet, if you're intent on batteries, is to regard them as another member of your family. They need attention, maintenance, their own living space, and so on. (They do not, however, require tuition, nor do they get married to somebody you loathe.)

## Lifetime

Batteries should last a few years when used properly. Expect about four years. If this is your first time with batteries, expect around three years. If you're like Curly, Moe, or Larry, expect one year or death, whichever comes first.

A battery monitor system can help you keep track of battery performance. At a cost of around $250, it's probably worth it. It can tell you exactly how much capacity your batteries have and how much they've been used over their lifetimes. No surprises.

## Risks

Batteries are full of nasty chemicals. Really nasty. And they produce poisonous gases, so it's not enough to simply avoid touching the chemicals. Sealed batteries avoid the gassing issue, but their performance isn't as good (meaning they're more expensive in the long run).

If possible, while you're using them, bury sealed batteries in a sealed container. Design this in as part of your system. Batteries can inflict much less mischief if they're buried underground. Avoid putting them in your house if you can — garages are good if they have ventilation. Always keep children away from batteries. When it comes time to discard the batteries, never bury them; take them to a recycling center. You will have to pay for this service, but please do it.

Risks aren't limited to toxic substances. Batteries are capable of outputting huge amounts of current, and this can be even more dangerous than an electrical shock. The voltages from a battery aren't generally high enough to cause electrical shocks (although if you connect enough batteries in a series, you can get a good shock). But a battery can discharge enough current to melt metal, creating a very serious burn hazard (machine shops and manufacturers use electricity in a similar way in arc welding). If you have a big battery bank, and drop a bicycle or a metal tool onto it, you may be treated to a Fourth of July spectacle right in your own house! Be careful!

# Specifying and Pricing Your Systems

At a minimum, you need to determine the following in order to get a good idea how much your standalone PV system will cost:

1. **Calculate the total watt-hours per day of energy you'll need.**

   Compile a list of all your appliances and devices and how many hours per day each will be run. You can find the wattage rating by either looking at the label on the appliance (almost all appliances will have a label) or making a measurement using a standard watt meter (available from Real Goods at www.realgoods.com).

2. **Define your peak instantaneous power output, measured in watts.**

   Determine which appliances you'll be running at the same time; add their power draws, in watts. You have a lot of control over this specification because you can run appliances at different times.

3. **Figure out the duty cycle.**

   *Duty cycle* is a measure of how often the system is used. A weekend cabin used for two days has a duty cycle of ⅔, or about 28 percent. A system used every day has a duty cycle of 100 percent.

**4. Estimate how many hours of good sunlight a day you can expect.**

Sunlight is difficult to estimate with much accuracy because it depends on the weather. It also depends on the time of year during which you're interested in using your system. For a weekend cabin in the mountains, used only in the summer, you may expect eight hours a day of sunshine, and in the thin mountain air, you'll get very good solar exposure. For a cabin used twice a month for a few days all year round, the worst case will be in winter, when you may only get three hours a day of cumulative sunshine, maybe even less if it's snowing and the panels are covered. Assume worst case, and be conservative on top of that.

In the following section, I give an example of how to put all this data together to come up with a final answer.

# Peeking In on a Real-Life Scenario

In this section, I walk through some of the calculations for an off-grid home in the mountains of Northern California. Here's what you need to keep in mind about this scenario:

- ✔ The duty cycle is 100 percent, and the house is used year-round.

- ✔ Worst-case expected sunlight per day is around four hours.

- ✔ A two-day power reserve is required because the backup generator is 20 years old and may or may not start, depending on its cantankerous mood. There may be some days when power is simply not available, but this is acceptable because the residents like to read, and a few good candles will work just fine.

- ✔ The system must output 120VAC.

And in case you're curious, here's how the owners are reducing their energy requirements:

- ✔ They use a wood-burning stove exclusively for heat.

- ✔ Residents don't need electric-powered water heating because they use a solar water heater.

- ✔ Both the cooking stove and refrigerator work with bottled propane.

✔ The house is extremely efficient, with well-designed window overhangs, a sunroom on the southern front, and a modular design that enables the living area to be closed off to the rest of the house on the cruelest winter days.

✔ A solar attic vent fan is installed in the attic space, and a large solar-powered ceiling fan in the great room keeps the comfort level on the hottest summer days tolerable.

## Calculating needs

Table 18-1 shows the sample power load for the California cabin (see Chapter 2 for a list of typical power draws for different household loads).

| Table 18-1 | Energy Consumption in an Off-Grid Home | | |
| --- | --- | --- | --- |
| *AC Device* | *Watts* | *Hours/Day* | *Watt-Hours/Day* |
| Kitchen lights | 120 | 6 | 720 |
| Family room light | 120 | 4 | 480 |
| TV | 70 | 3 | 210 |
| Coffee pot | 200 | 0.5 | 100 |
| Clock radio | 1 | 24 | 24 |
| Table fan | 15 | 6 | 90 |
| Computer | 100 | 7 | 700 |
| Miscellaneous appliances | 400 | 0.5 | 200 |

With this chart, you can do some of the following calculations for your load analysis. To calculate the last column, simply multiply the first two columns.

| *Calculation* | *Answer* |
| --- | --- |
| Total energy needs in one day | 2.5 kWh/day |
| Adjustment for inefficiency (10 percent) | 2.8 kWh/day |
| Maximum instantaneous load | 700 W |
| Duty cycle | 100% |
| Total energy needs in one week | 19.4 kWh/week |
| Power needed from PV panels (@4hrs/day) | 700 W |

Next, calculate battery size, which is specified in terms of amp-hours (Ah). Most batteries are 12VDC, but other sizes are also available. For this example, assume a 12VDC system.

1. **Take the total kWh/day, multiply this by 1,000 to get kWh/day, and then divide this value by the battery voltage.**

   For the example, this yields 233Ah:

   $$\frac{2.5 \text{ kWh/day} \times 1,000 \text{ W/kWh}}{12\text{VDC}} \approx 233 \text{ Ah}$$

2. **The generator is old, so triple the result from Step 1 to account for the two reserve days.**

   To cover the two reserve days, the cabin owners need batteries that can hold three days' worth of charge, which yields around 700 Ah:

   $$233 \text{ Ah} \times 3 \approx 700 \text{ Ah}$$

3. **Multiply the minimum battery capacity from Step 2 by a factor of two.**

   Batteries last much longer when they're not drained of more than about 70 percent of their available energy. In the long term, getting a battery with a larger capacity will cost much less than replacing batteries more often — plus a performance margin is always nice (the net effect will be more reserve days). Therefore, the residents need a total of 1,400 Ah of battery capacity:

   $$700 \text{ Ah} \times 2 = 1,400 \text{ Ah}$$

Finally, they're going to need a larger PV module capacity in order to get the three-day reserve. It's okay to go without power a few times, and there's a backup generator, so if they double the size of the PV module capacity, they should be safe. Hence, the owners need 1,400 watts of PV.

## Checking out the complete system

Take a look at the components of the complete system. You can install a system like this yourself for around $14,000. In many off-grid applications, you don't need to get a county permit or have inspectors look at your system. You're pretty much on your own, although you should be aware of your local requirements.

| Quantity | Device |
|---|---|
| 10 | 140W PV modules |
| 1 | Integrated power control system (charge controller, inverter, switches, and so on) |
| 1 | 30-amp two-pole safety disconnect with fuses — 1,500 W total |
| 1 | Rack-mount system for 10 PV modules (adjustable for season) |
| As required | Wiring |
| 8 | 183Ah, 12VDC gel storage batteries with interconnect cables |
| 1 | Plywood box for safely containing the batteries |

*Note:* Peace signs are optional, although for some reason, they seem to go hand in hand with off-grid lifestyles.

Sealed gel storage batteries are used because they're safer; they cost more, but it's worth it, especially if kids are around.

In your system, you need to aim the PV modules optimally (see Chapter 5). You may want to use adjustable mounts so you can get more out of the system in the winter by aiming the panels lower in the sky. If so, mount the PV panels on the ground, instead of a roof — you can make adjustments easily and often.

Using wind or water doesn't change things much. Just substitute a hydro generator or wind turbine for the PV panels. That energy source may be much less reliable, but on the other hand, it may be more reliable, particularly with water power. You can also use a combination, which is what a lot of off-grid home styles incorporate. (See Chapter 14 for more on wind and water.)

Remember that when you do your calculations, use conservative estimates and allow for worst-case scenarios. Over the course of the summer, and when it's sunny and clear, the system has much more capacity, maybe double. And the need for reserve is less because the odds of needing it are much less.

Ironically, the power consumption is less in the summer because the days are longer and the residents are living outside more. But that's the nature of a solar system. You don't save anything by using a solar system at less than capacity, so the net result in practical terms is that in sunny weather, you can use more power without worrying about conserving. For instance, you can watch more television.

# Chapter 19

# Do It Yourself or Contract Out?

*Y*ou can do all the work on any project yourself. But do you want to? Or should you? You can potentially save a lot of money by doing the work on your own, but you can also run into problems and end up paying a lot more to get things fixed when your system doesn't pan out the way you expected.

In this chapter I give you a variety of factors to consider when making the decision to either do a project yourself or hire a contractor. Also keep in mind that you can usually hire a contractor to do only a portion of a job for you. At the very least, you can buy a kit that you can install yourself, and this has the advantage of being well designed and proven. (In Part IV, I address large-scale solar PV projects that you can choose to do yourself or contract out to a professional.)

# Design and Installation Issues That Can Influence Your Decision

First, you need to understand that a good solar PV design entails a lot of small details. Here's a fact: Two identical systems in terms of parts and equipment may operate very differently. If you use wire sizes that aren't just right, you lose production, which translates into a poor return on investment. In addition, a solar PV system has the potential to be very dangerous. The voltages are high, the power levels are high and the system is going to be installed outdoors. Details are often subtle and counterintuitive.

In addition, weather and sun exposure create difficult engineering problems over the long term. Sunlight is corrosive and will degrade the quality of metals and plastics after years of exposure. In particular, electrical connections need to be made with the utmost attention to integrity. When weather (rain) constantly barrages connections, all kinds of nasty things can happen, none to your advantage. The point is this; you may install a perfectly good solar PV system today, only to discover that in a few years things start to decay, and the system may stop working properly or, even worse, become dangerous.

Bottom line: Experience counts for a lot. It's not just a question of understanding the technical ins and outs of a project; it's also important to understand how things can go wrong over time. A good contractor understands these things, and knows how to prevent them. After all, he's going to be offering you a warranty (usually 10 years for the entire system, and 25 years for the solar panels). When things go wrong, it costs the contractor a lot of money, not to mention the cost to his reputation.

For the vast majority of homeowners, a contractor is the best choice. If, just reading this section, you've made up your mind to hire a contractor, you can skip straight to the section "Hiring a Contractor." If you're still on the fence or have decided to go it alone, you'll need to be well versed in tools, plus you need to understand electrical fundamentals well enough so that you don't hurt yourself or anybody who may come in contact with your solar PV system. The following sections explain what else you need to know and where you can go to get the necessary details.

# Getting Good Information

If you're thinking about doing a big project at your house, understanding the details of the entire process — from design to completion and use — is by far the most important place to begin. Take your time; don't set a deadline or a schedule.

Here are some important steps to take:

1. **Research your subject (which you've already started doing by reading this excellent, well-written book).**

   Go online and find information on every system you can that looks like the one you want. Look up manufacturers — how many are there? Are they new? Are there alternatives? In particular, you want to look for kits whenever possible, regardless of whether you're doing the work yourself or contracting.

2. **Make drawings, including parts lists and specifications.**

   Don't worry if they're not completely accurate; just making drawings can move you along in the right direction. If you can, obtain some formal drawings (those that have been approved by a county building department) and make your own drawings look the same. The more complete you can make your drawings, the easier your project will go.

3. **Take a trip to your county building department and informally inquire about what you're going to do.**

   The best resource you can obtain is a friendly, courteous, willing clerk. Good luck! Show the staff your informal drawings. These people are bureaucrats, but be nice, cooperative, and patient. Don't get angry, no matter what they say or do. They can offer some really good advice, and at the very least, you can find out what you're going to be up against if you decide to enter the formal permit process on your own.

   The upcoming "Passing code" section explains questions you should ask during this trip.

4. **Call around.**

   Call contractors, trade agencies, and stores that sell equipment. Ask them about the system you are contemplating; ask them the pros and cons and what advice they can offer. Always talk to more than one resource so that you're not led in a particular direction that they want you to go (the direction that will make them the most money, of course).

5. **Contact your utility and discuss the project with them.**

   Your utility will probably try to talk you out of doing it yourself. Most don't like customer installed PV systems, and for good reason. You may not even be able to install your own system; your area may have a law forbidding laymen from installing. If so, you can still manage to do a lot of the work yourself; you just need to find a contractor who will let you do the work and then sign off on it after he or she has ascertained your system's integrity.

# Being Realistic about Codes and Regulations

Every major project on your house is subject to a whole book full of codes and regulations. In theory, government regulations exist to help people maintain safe, reliable, predictable, and consistent systems, but the permit system doesn't always work as planned. Even if you're in the business, trying to stay abreast of all the codes, regulations, laws, fees, and so on regarding the solar industry is prohibitive. In the following sections, I commiserate with you about the aggravations, note some of the reasons you should try to follow code, and give you tips to help you complete the process successfully.

## On the honor system: Choosing to follow code

Red tape is a major hassle. Bureaucrats often seem more concerned that you've dotted your *i*'s and crossed your *t*'s than they are with safety and reliability, and applying for permits can easily double the complexity of a building project (in terms of your labor time). Your drawings themselves must follow code — which can be difficult to interpret and implement — and if you make adjustments to your original design, you have to resubmit your drawings and get new permits. Even for experienced solar contractors, it's not untypical to visit the county building department several times prior to final approval.

Code basically says you shouldn't rewire anything in your house without a permit and an inspection. The fact is, though, you may be able to install a solar system completely on your own — it'll likely work well, give you the performance you anticipated, and be safe and reliable. Perhaps nobody will find out you didn't get permits, and when you go to sell your house, the subject most likely won't even come up. And even if it does, most buyers won't flinch because most houses have work that's not to code. Homeowner's insurance policies often contain clauses that cover subcode (or uninspected) installations on your house. But ask yourself this question: Would you buy a house that had a solar PV system installed by a layman, without permits? I wouldn't.

So if wading through the red tape is a big overhead rate that provides little benefit, why bother? Here are the some of the biggest reasons:

- ✔ **It's the law.** Codes are laws, and I can't advocate breaking the law. (And if you do build without a permit and a vindictive neighbor turns you in, you may have to pay fines, stop work, and/or remove the work you've already completed.)

- ✔ **Codes provide safety standards.** Especially if electrical or plumbing work is new to you, codes can inform you of safety issues. Following code can also help ensure that you don't knock out those pesky load-bearing walls or install a collector your roof can't support. When you install solar PV panels on your roof, for example, they must be able to withstand winds up to 125 mph. (A panel that comes loose and blows into the neighborhood would be one massive Frisbee that could easily decapitate somebody. Yuck!)

- ✔ **Contractors can get into legal problems for not getting permits, but even more to the point, it'll be impossible for you to sue them if something goes wrong.** Courts don't recognize your legal right to sue if you're working outside the system, and rightly so. (Contractors can take care of the permit and inspection process for you — see the upcoming "Hiring a Contractor" section.)

✔ **Following codes and getting formal permits reduces the legal liability you may face if somebody is injured on your property.** If you want to play it safe, go the formal route. You may want to consult with your insurance agent to find out how your liability insurance is affected if you have noncode work. In fact, you may want to consult with your insurance agent about installing a solar PV system because they may increase your rates if you do. Why? There are added liability issues associated with PV, and some insurers don't want to deal with these. Time to find a new insurance company.

✔ **Many rebates and warranties require you to go through the formal permit process.** You can't really blame manufacturers for insisting that the equipment they're selling you is professionally installed because they take on more risk if it's not. In fact, there are some manufacturers who won't sell equipment to laymen. And government agencies want code-compliant work because, well, they're government agencies and they support all other government agencies.

## Passing code

Final permit applications and drawings should be as polished as your resume or an important business proposal. Here are some tips that can help you make sure that your drawings and installation pass code:

✔ The best way to find out about codes is to visit your county building department. Tell the inspector exactly what you're planning to do and ask him what you need to do to comply.

✔ Base your drawings off of formal drawings that have already passed code. You can find these drawings from friends or from the building department at your county government.

✔ Informally visit the county building office with your initial drawings and ask about them. Also ask as many questions about the permit process as you can, including the following:

- How long is the wait?

- How many forms are required?

- How long does it take to fill out the forms?

- What are the fees?

- What inspections are required?

- What happens if you do a job yourself and it doesn't pass inspection?

- How do you request an inspection?

✔ Make your drawings as complete as possible, and include detailed materials and spec lists. Make copies of all the specification sheets for all of the parts and equipment you plan on using. Bureaucrats often reject drawings for not having the proper legends, so make sure that yours are accurate. Take your plans to the building department and be prepared to go through another round when it looks at them and tells you that you need more work. It's unlikely you'll get your plans approved the first time through the mill. That's okay, because above all, you want your plans as good as you can get them. By using the inspectors as information resources, you'll end up with a better final product.

# Hiring a Contractor

Part of your decision on whether to do a project yourself or hire a contractor should involve understanding exactly what a contractor can do for you and what the process will be. Be aware that after you sign a contract, backing out and going it alone is going to be difficult. In the following sections, I describe what a contractor does and how it affects your project timeline.

## Knowing what a contractor can do

Every contractor uses a standard contract and follows approved practices. You always have the right to negotiate; contractors always have the right to refuse. You can engage a contractor at any point in the project, but most are very reluctant to come in after you've done half the job (or if they do, they're inclined to charge a lot and smirk at your ignorance). Most want to be a part of the design, which may or may not cause problems. You can hire a contractor to do an entire project, or just parts of it.

Here's an interesting angle to consider. Contractors buy equipment and parts in bulk, so they can get a better price than you (theoretically). It may turn out that they can do the job for less than you can do it yourself simply because they can buy equipment so much cheaper than you can. By the time they add their labor expenses, the difference is still in their favor. While it may be compelling to do a job yourself, due to the personal satisfaction, it may actually cost you more.

You can save money by negotiating to do any of these items yourself. You can do a design and find a contractor who will build accordingly. Or you can be responsible for keeping the work site clean, and you can probably come to an agreement that you assist on the job as much as possible. You may even find a contractor who'll let you do all the actual work but who will do the drawings, permits, and inspections for you.

# Getting bids

Always, always, always get multiple bids. Even if you're planning on doing a project completely on your own, you should get bids. (I can hear from my contractor friends that this wastes their time and that they resent it, but you can get a lot of good, free information this way.) Let contractors know you're going out for competitive bids, and never make a commitment of any kind. Many contractors offer a lower price, if you sign the contract *today*. That's bunk; don't fall for it. If you call them back in a couple weeks and ask for the same discounted price, they'll give it to you.

The best bet is to ask friends who have the same kind of system, you're contemplating for contractor referrals (or warnings to stay away, as the case may be). You can look contractors up with the Better Business Bureau, or you can get information from state regulatory agencies. The Internet contains a lot of referral sites (but beware — these may be paid for by the contractors themselves!). Type the name of your city or county and the kind of project you want to do into your search engine.

Never use family members or friends as contractors. You may think they'll give you a better price and better service, but you'll probably regret such an arrangement. They may be thinking they can charge you more because of the lack of competitive bids, or that they can work your job when it's convenient because you won't fire them. Of course, if Uncle Roy the solar installer discovers you had a solar installation done by one of his competitors, he may not show up for Christmas dinner, so tread carefully. (On the other hand, maybe you don't want Uncle Roy around for Christmas dinner because he blows his nose every couple minutes and grosses everybody out.)

# Comparing bids

After you get the bids, the problem becomes one of comparing apples to apples, which is almost always difficult with solar because the industry is fairly new and there are so many different ways to do things. Choosing the contractor comes down to three criteria, which I explain in the following sections.

## Craftsmanship

To evaluate craftmanship, get referrals. Go to a site where the contractor has completed a job and look at the solder joints, the weld joints, and the neatness and simplicity of the installation. How long did it take? (Although time may not seem important, it reveals experience and efficiency.) How long has the contractor been in the business? Look at older installations as well as the most recent.

If you can, find a job your potential contractors did that they don't mention on their lists of references — then talk to the owner. That's where you can find some real information. All contractors can give you a list of happy references and show you a pile of after-job surveys that are glowing and exuberant. But what of the customers who aren't on these lists? Ask contractors how many jobs they've completed and then ask why there are so many fewer references. (They won't like this question!) If they can't supply a list of references, don't use that contractor.

Make sure to ask about workmanship warranties. If contractors are offering only the warranty that comes from the manufacturer, beware. Ask about bonding, insurance, licenses, and so on. If your contractor doesn't like these questions, watch out. Good contractors are glad to hear these questions; it means their shady competitors will be weeded out.

### Compatibility

Some contractors are just dubious characters, and some are living in a different universe than you and I. The bottom line is that you want a contractor who, when he sees your phone number on his caller ID system, is happy to answer the phone.

Having a good relationship matters a lot. Not only does it foster cooperation and increase the likelihood that problems will be ironed out quickly and effectively, but a contractor who doesn't like you simply isn't going to go the extra mile for you (that's right — the feelings can go both ways). You may get the work done as per spec, but you won't get the extra time spent to get the connections just right. Or you won't get your phone calls returned until it can't be put off anymore.

### Cost

Cost is very important, but ultimately, it may be one of the least significant factors. Why? So many things can go wrong, and problems generally add up to more cost. If you get a bid that's much lower than the other ones, beware. Your contractor probably doesn't understand what he's doing, either technically or competitively. Or maybe you're getting a contractor with a much lower overhead rate. Why? Does he not carry insurance? Is he working out of his truck or off of his bicycle? Will he be around in three months to provide warranty coverage?

Still, you can find very low costs that work out truly well for you. The industry is growing rapidly, and large companies are interested in expanding their markets, and so they charge super-low prices in the new regions they're entering just so they can establish their presence (and push out the existing contractors). In other words, they'll sign up to take some losses in order to create more business and profits in the long term.

Here's some advice. Negotiate! There are no rules against it. Once you have your bids, call the various contractors and ask for a best and final price. Chances are, you'll find that a contractor will cave in a little in order to get the job. The business is highly competitive; use that to your advantage.

## Interviewing a contractor

Contractors all want the job for the most amount of money they can get. Here's how to wade through the quicksand. First, let the contractor do the talking during the interview. If you're talking, you're not getting information. If you feel the need to convince a contractor to do a job, you're setting yourself up for a disappointment down the line because these folks are generally good at smelling blood. Next, ask the same questions of each contractor. Ask each the following:

✔ **What problems they foresee.** If they say none at all, beware.

Give a hypothetical scenario such as, "What if you're halfway through and find that the parts aren't fitting together the way they're supposed to?" Or "What if you're injured somehow?"

✔ **Whether they use multiple sources of supply.** If they're stuck on one single system supplier, ask them why and whether they'd install a system from another supplier if you were to purchase the material yourself. Don't exclude them if they refuse; just find out why.

✔ **Whether they've used the same brands of equipment for a number of years.** If they've switched often, ask why. It's normal for contractors to switch equipment; the industry is changing rapidly. So don't exclude a contractor if they've switched suppliers. Just find out why.

✔ **Whether they're committed to finishing the project after they begin.** Be clear upfront that finishing the project on time is one of your requirements. Ask them what can be done if a project stalls.

✔ **What conditions would merit tossing out the contract altogether.** For instance, if the equipment they bid is no longer available, do you have the option to cancel the contract, or do you need to accept their replacement equipment?

✔ **Who will be doing what.** Who will design the job? Who will oversee the project management? Who does the installations? Get a list of all the people who will be involved and their phone numbers.

# Contracting

When you sign the papers, let your contractor know that you're very happy to have him doing your job. Tell them that the process was competitive, but that they outshone all the rest (in some circles this is called "sucking up," but we don't use that kind of language in a *For Dummies* book). Begin working with your contractor well before your contractor begins work on your project.

When contracting, get everything in writing. And throughout the entire project, clearly tell the contractors that everything will be in writing. Any decisions made or changes made will be in writing only. E-mail is okay, but print out e-mails and keep them in a file folder. I repeat: Get everything in writing. Get everything in writing, all the time, always. It's not that people are to be distrusted, but verbal communication is like that children's game "Telephone," in which the message gets garbled each time it's spoken. Writing forces clearer articulation.

In addition to a lot of verbiage and conditions that are generally boilerplate, a contract should include the following:

- Prices, preferably broken down into line items (but don't insist)
- A schedule of verifiable events that the two of you can easily establish and approve
- Cash flow (when payments are due, how much they are, and so on)

Be aware that contracts must be legal. In other words, neither you nor the contractor can agree to do something that's illegal. Also, a contract cannot violate standard business practices, so something in the contract may simply be unenforceable by the courts.

Never pay a contractor until the work is finished. The best bet is not to make any payments at all until the entire job is finished. You may need to make progress payments, but these need to be very well specified. For example, you may pay 20 percent of the contract amount once the system design and site survey is finished. You pay another 20 percent when the panels are installed. Another 20 percent comes when the system is connected to the inverter, and the final payment comes when the final inspection is approved. In most states, it's illegal for a contractor to precharge for work, aside from a modest downpayment to get the contract initiated. Go to Chapter 16 for more details on working with a contractor.

# *Working with a contractor after the job begins*

When you're finally on the go, you need to work with your contractor as effectively as possible. Now is not the time to express doubt about her abilities. Now's the time to follow the Golden Rule: Do unto others as you would have them do unto you:

- ✔ Let her know that you think this is a two-way street. Of course she has responsibilities, but so do you. When she knows that you understand this, she'll be more enthusiastic.

- ✔ Always be completely square and decisive. You may not want to express disappointment about how a job is turning out, but you had better if that's what you're feeling. Let your contractor know that that's what you expect of her as well. In my experience, playing word games always leads to problems.

- ✔ Communicate routinely. Every morning, before work starts, go over the project. Ask about problems. Ask about the schedule and the budget. Ask, "What do you need from me?" Even if it's only for a few minutes, establishing the routine is worth the effort.

- ✔ Be friendly no matter what. Don't get emotional. Problems happen; be cool when they do, and your contractor will be more honest with you about everything.

- ✔ Offer cold drinks. Make a cooler available. Go the extra mile.

- ✔ Don't forget the need for a restroom. Make it nice and easy to get to, and keep it clean.

- ✔ Establish some ground rules that both of you must follow. When is it okay to communicate? When shouldn't you call? Who needs to keep what clean? Should workers be allowed to use radios or remove their shirts when it's hot? Can they hose themselves down to cool off? Ask them whether they have any ground rules they would like you to follow.

# Chapter 20

# Finding the Cash to Get the Job Done

. . . . . . . . . . . . . . . . . . . . . . . . . . . . . . . . . . . . . . . . . . . . . . . .

## In This Chapter

▶ Looking in to the various subsidy options

▶ Using loans to finance your solar projects

▶ Working with banks and financial institutions

▶ Doing research to find out more

. . . . . . . . . . . . . . . . . . . . . . . . . . . . . . . . . . . . . . . . . . . . . . . .

The fact is, solar power is not a cut-and-dried winner in terms of investments. (Chapter 6 details payback analysis techniques that you can use to calculate the value of a solar investment to compare it with other types of investments.) The bottom line for a PV solar investment depends on three things in particular: cost of energy (which will rise); amount of sunlight available (which varies greatly from region to region — see Chapter 5); and cost of solar equipment.

Although basic equipment costs don't vary much between regions, one thing that does affect how much you pay for your system by region is government subsidies. This chapter details the basic ideas behind subsidies and then shows you how to do your own research to determine how best to obtain and use them.

This chapter also covers the various financing options and offers sage advice on how best to proceed with the banks and financial institutions that offer the loans.

# The Different Types of Subsidies

Governments have recognized the value of solar energy, especially in achieving a more pollution-free energy consumption. Plus, solar power has the potential to free North America from reliance on foreign oil, which is desirable from a political stability standpoint. (Getting oil from the Mideast creates a lot of problems.) Governments also recognize the marginal nature of a solar investment, and they've resolved to do something on the cost end of the equation (some politicians may claim to be able to do something on the sunshine supply end, but as usual they're just blowing hot air).

*Subsidies,* where the government helps you pay for part of the cost of your solar system, are very powerful incentives to invest in solar, and a solar customer can see up to a 50-percent cost reduction by using all the subsidies available. However, the subsidies scene is constantly changing, so it's impossible to specify the precise details of subsidies and rebates simply because they're apt to change without notice. I list a number of resources for you to consult to find the timely information that you will need.

The following sections describe the types of subsidies available.

# Rebates

A *rebate* is money given back to a customer after he or she makes a purchase. Manufacturer's rebates work the same as they do for other products. You buy a solar system or component, and the manufacturer gives you either an instant rebate at the checkout or one that you need to mail in with proof-of-purchase.

Government solar rebates operate basically the same way. The rebate is paid directly to the professional solar contractor, so the customer doesn't need to finance the entire investment and then wait to get the rebate back. This is, in effect, a secondary rebate because it allows solar customers to more easily purchase their systems. But because these rebates come from government sources, they're also laden with overwrought details generally incomprehensible to the average layperson.

For example, here's how rebates work in California. For a PV system, you can get up to 25 percent of the cost of a system in rebates, but the amount depends on the system's productivity. So the state has a very specific computer program that potential customers use to predict system performance. Inputs are the following:

✔ **Your geographical area:** Data banks show the statistical weather patterns in your area. This includes cloud density and frequency, average temperatures, air density, and the like. Your latitude also gives an indication of sunlight exposure. All these numbers combined yield a measure of how much useable sunlight you can expect in a given year. Some *microclimates* (a climate that, because of terrain, landscaping, and other factors, may differ from a climate only a mile away) may have much less sunshine than a nearby area, so be prepared for some surprises here.

✔ **The size and orientation of your solar collectors:** The greatest solar exposure comes from a specific roof angle (depending on your latitude) and a specific azimuth angle, namely true south (see Chapter 5 for details). For any angles other than these, you get a lower rebate. If your roof is all wrong, not only do you get a lower rebate, but you also get less productivity as well.

To maximize your rebate, you may be able to mount your solar collectors on the ground and get all the angles just perfect. You can also get a higher rebate by employing adjustable mounts that track the sun over the course of a year — and result in greater productivity.

✔ **Any shading issues inherent to the collector location:** A special hemispheric reflector device is used to measure the relative location and shading effects of trees, mountains, chimneys, other buildings, and so on. The device can tell whether these impediments will affect your solar exposure at any time of the year. A computer program translates this data into a single number that tells how much solar potential your collector site has.

✔ **The efficiency of your system:** This is also related to a system's productivity. Manufacturers must define a system's efficiency, but this number also depends on where the collectors are located. (If they're on the northern side of your roof, the score goes down.) The overall system productivity is what's of interest; this is the total amount of energy a system will produce in a given year. The government is subsidizing energy production, not a system. It's a fine point of distinction, but it makes sense. (As a matter of interest, in California, you may choose to get your subsidy after the system is installed instead of using forms and paperwork to calculate a theoretical efficiency when the system is installed. In this way, the actual performance is used to gauge how big the subsidy will be, which often results in a larger subsidy than the necessarily conservative theoretical.)

For a perfect solar location, the state rebates up to 25 percent of the cost of your solar system (acceptable market costs are specified so that you can't get a higher rebate for a system that's out of market pricing guidelines). In actuality, 25 percent is rare. Most numbers come in around 20 to 24 percent. When they come in much lower, your solar investment quickly loses allure.

You may run into a *buy-down* caveat: As time progresses, the rebates are scheduled to decrease, which encourages solar customers to buy now instead of later. The reason for this is reasonable: The basic idea with subsidies is that they'll force the cost of solar to come down over time, so why invest now? Why not just wait until prices do come down? But if everybody were to have this philosophy, prices would never drop. Ideally, as the rebates diminish, prices will come down due to market forces, so the net cost of a system won't change over time. (Well, okay, but when do government programs ever work the way they're supposed to?)

The effect of this rebate has been wildly popular in California, which has some of the highest power rates in the country, plus the unfortunate prospect of experiencing greater increases in energy rates than the rest of the country due to the robust environmental laws that the state legislature is constantly passing. In addition, Californians remember (not fondly) the energy price spikes that resulted in companies like Enron playing with energy prices, resulting in higher profits for the company at the expense of the energy consumers. There were blackouts and brownouts and power outages in the middle of steamy, sunny afternoons.

## Tax credits

*Tax credits,* which are subtracted from the amount you owe on your state or federal taxes, are an increasingly common form of subsidy. You must buy a certain type of system to qualify, and the credit is taken on your tax return, either state or federal. As a result, you must carry the cost until your return is filed, and any refund comes back to you. This wait can take more than a year, depending on when you buy a system, and when you file your taxes.

In the fall of 2008, as part of the stimulus bailout bill, the federal government set the Investment Tax Credit (ITC) for solar at 30 percent, without a cap. Prior to this, there was a cap of $2,000. This has made a huge difference for customers, needless to say.

A tax credit is far better than a tax deduction because a *tax deduction* is subtracted from your taxable income. So if you're paying a marginal tax rate of 28 percent and you get a deduction of $1,000, you save $280. But tax credits come out of the amount you owe, so a $10,000 tax credit saves you $10,000.

Both tax deductions and tax credits may be available for a solar system. Many times, both the federal and state governments are offering programs, and the terms will vary. It's important to understand the distinction between credits and deductions because they make a big difference on the bottom line.

Here's an example. The U.S. federal government allows a 30-percent tax credit for PV systems, with no cap on system size. If your solar PV system costs $30,000, and you get a state rebate of $5,000, the net cost is $25,000. You get a tax credit of 30 percent of this, or $7,500. So your solar system now only costs you a net of $17,500. In essence, the government is subsidizing 41.6 percent of the total system cost.

## No property tax increases

Many states have laws that prevent your property taxes from rising due to the increase in value of your home from a solar investment. So if you install a $40,000 PV system, your county can't reassess your property and charge you extra taxes for having a home that's worth $40,000 more. Very few other home improvement investments qualify for this exemption, and because it's basically being paid for by the government, it's a tax break.

## Incentives for home-operated businesses

The incentives for businesses are even greater than those for residences, so if you have a home business or office, you may qualify for higher tax credits and rebates. Why? One reason is that businesses use most of their energy in the middle of the afternoon, so they're less apt to go for solar energy when there's net metering. Another is that the green spirit that motivates many solar investments (thereby making the investment more attractive than mere numbers would imply because you save pollution as well as money) doesn't exist for businesses. (For a way to put a dollar value on pollution savings, see Chapter 6.)

Businesses can also take advantage of accelerated depreciation schedules for solar equipment. This stuff can get complex; ask your tax preparer for the details. (Ugh! Now I sound like one of those irritating mortgage-broker commercials on the radio.)

## Net metering

Utilities aren't particularly enthusiastic about offering rebates, nor do they really want residential customers to hook solar equipment up to the grid because they lose revenue. Recognizing this, the 1978 federal Public Utility Regulatory Policy Act (PURPA) mandates local utilities to pay "avoided" or wholesale costs to entities that want to sell it. This is the basis for net metering, which makes solar PV investment very viable (see Chapter 17 for details of the whys and wherefores). Without net metering, solar PV would be no more than a rare novelty.

States often have even more stringent net metering rules. In California, the utilities must pay you the same rate that they charge for power. For example, if you're on a TOU (time of use) rate schedule, the utility must pay you the same rates it's charging its customers at different times of the day. This arrangement is a very strong incentive for using solar power because the highest power rates always apply in the middle of the afternoon, during the summer months. Midafternoon is the highest demand period because businesses are all using air conditioners at the same time, but it's also the time when solar PV systems are outputting at the maximum rate. So if you have a PV intertie system, you can plan your power consumption to take advantage of the non-symmetric exchange of energy (see Chapter 17 for the technical details).

## Tax-deductible home equity loans

Although they may not be a direct subsidy of solar power systems, tax-deductible home equity loans are akin to a government subsidy. If you take out a second mortgage to pay for your solar PV equipment, the government lets you deduct the interest on that loan from your taxes. So if you're making a monthly payment of, say, $300 for a second mortgage, your net cost may be only $200. The tax-deductible home equity loan works exactly like a subsidy, but it's called something else.

# Researching All the Subsidy Options

Things are changing so fast that it's impossible to keep up with all the government programs. I can only point you in the right direction and give you a nudge. Your job is to poke around and get all the details for yourself. Here's how to research the subject further:

- **Talk to your tax preparer.** Tax preparers should be versed in the details of solar tax credits, but if not, they can still access the information sources they need to process your rebates and tax credits. Solar is relatively new, so give preparers a break if they don't know the details upfront, but keep at them to find out if they don't.

- **Ask contractors.** Solar contractors need to know about subsidies because they help sell systems. PV contractors are experts at tax credits and rebates, and they'll usually help you process them as well. Even if you're going to do the installation yourself, you can get good information from contractors when you ask them for competitive quotations.

- **Contact utility companies.** Utilities will give you information, although they don't necessarily want to. But they have to (it's required by law), so if you push a little, you can get all kinds of information, and it's almost always free. You can find a customer service number on your utility bill. When you call, ask if a particular department is dedicated to subsidies and rebates.

- **Search online.** Of course, the Internet is a great source of information. Use a search engine to look up key words like "solar energy tax incentives," "solar power rebates," and so on. One thing will lead to another, and who knows where you'll end up?

# Getting a Loan For Your Solar PV System

When you apply for a loan, regardless of where you apply, the main factor the bank is interested in is *risk,* which is the likelihood that you'll default on your payments or perpetually make late payments. When risk is high, the loan's interest rate is high; it's very simple. If you want to take out a loan for a solar project, your job is to convince the financial institution that you constitute low risk. In this section, I show you how to increase your odds of successfully obtaining a good loan (that is to say, one with a low interest rate).

The government is interested in promoting solar power, so it has a wide range of loan subsidies available. In essence, the government accepts part of the risk of a loan, so a bank can offer you a lower interest rate. It's worth your while to spend some time looking into the various government programs because you can often get a loan for better terms.

## Borrowing money the old-fashioned way

When people need large amounts of money, they either look to a bank or whip out their credit cards. To finance your big solar projects, you can pursue several types of loans:

- ✔ **Home equity loans** are the most common loans and the best options. If you're contemplating a large solar project, you probably own your own home. (Why would anybody install a big solar system on a home she's renting?)

  In parts of the country where real estate values are rising, homeowners enjoy *equity,* which is the difference between what's owed in mortgages on the home and the home's market value (or appraised value, which should be the same but rarely is). You can use your equity to get a loan at a much lower interest rate because the bank's risk is greatly reduced by using real estate as *collateral* (the goods the bank will take if you don't pay on your loan). The bank won't need to come after the equipment you've financed because the bank can go after the house itself (banks generally don't care what you do with the money, so they may not even know about your solar equipment). Homeowners plan to pay off their debt long before they let a bank take their home away. Home equity loans are generally tax deductible (consult your tax preparer for details).

✔ **Supplier loans** are available from manufacturers and suppliers who provide solar equipment. However, solar vendors generally sign up with an equity loan broker and simply act as the sales outlet for somebody else's loans. You can probably find a better interest rate if you go directly to a bank, but the convenience may offset the cost. Or solar vendors may actually offer better terms because they have a strong incentive to close a sale, and making inexpensive financing available helps considerably.

These companies use the equipment itself as collateral for the loan. Supplier loans are similar to automobile loans that dealerships offer when you buy a vehicle. The car is collateral, and if you don't pay the loan, a repo man (or woman) comes to your house late at night and takes back the car — perfectly legal. If you default on a supplier loan for your solar equipment, the company may send someone in the middle of the night to grab the collectors off your roof — you probably won't mistake him for Santa Claus!

✔ **Consumer loans** don't require any collateral, so the lender's risk is high. You can get a credit card with a $30,000 credit limit far more easily than should ever be possible. The reason it's so easy? The interest rate is sky high; the bank expects a number of defaults and lets the customers who don't default cover the losses from the ones who do. Plus, you don't get any sort of tax deduction on the interest.

Consumer loans are the modern version of loan shark products, although the creditors no longer break your fingers when you default. If you're a good credit risk, avoid consumer loans like the plague.

✔ **New first mortgages** often make sense. Interest rates are very low right now, and if you have some equity in your home, you may be able to obtain a new first mortgage that incorporates the cost of your solar PV system, and the payments may not be any bigger than what you're making now (since the interest rate is lower). In essence, you'll be saving on your monthly utility bills, and your mortgage will be the same, so you're monthly cash flow will be much better. Now you can buy that new car you've been dreaming about. This is, among the many alternatives, the best way to go, but it all depends on the level of equity you have in your home.

## Using an energy-efficient financing program

A less-traditional option of financing your major solar project is through an energy-efficient financing program, but your home has to qualify for the program. To qualify, you need to have your home audited and rated by a licensed expert. He or she will do an energy audit (see Chapter 2) and write a report that estimates annual energy use and costs. To get the best loan terms, you have to convince the financing institution that the improvements you plan to fund with the loan proceeds make sense in the grand scheme of things. That's reasonable with a PV system.

If things look good, you can get special energy-efficient financing programs that have lower interest rates than conventional loans. Keep in mind, however, that with these loans, you'll be required to pay for the energy audit and deal with government agencies.

Residential Energy Services Network (RESNET) is a network of mortgage bankers, builders, and others. RESNET has solar loan programs available for the right customers. Check them out at www.natresnet.org.

## Pursuing energy efficient mortgages

There are two types of *energy efficient mortgages* (EEMs): those for either new or existing homes. You can obtain one of these loans for a new home, an existing home, or a refinance. The advantage is a lower interest rate, which is subsidized by the federal government.

Here's a quick breakdown of who's providing what, but keep in mind that I'm talking about the government, so terms and programs are subject to change at a lawmaker's whim:

- **Fannie Mae:** The Federal National Mortgage Association encourages lenders to provide EEMs by establishing guidelines and certain types of incentives. Check out what's available at www.fanniemae.com. You can also get a list of lenders who have signed on with the programs.

- **Freddie Mac:** The Federal Home Mortgage Loan Corporation does the same basic things as Fannie Mae but is generally more interested in long-term loans. Check out its offerings at www.freddiemac.com.

- **Farmer Mac:** This agency is part of the U.S. Department of Agriculture and is for those of you who are down on the farm. Go to www.usda.gov for details.

## Thinking about other mortgage options

Here are some other potentially useful government agencies:

- Department of Energy (DOE) at www.doe.gov.

- Department of Housing and Urban Development (HUD) at www.hud.gov.

- Department of Veterans Affairs (VA) at www.va.gov/vas/loan.

State agencies often provide subsidized loans as well. Check out the prospects at `www.naseo.org`, the Web site for the National Association of State Energy Officials.

# Considering Alternative Financing

You also may want to look at some other financing programs. In the sections that follow, I go over a couple of your options: leasing equipment or buying remote parcels of land so you can generate power and sell it back to the power company.

## Leasing solar equipment

In one new idea that's taking hold, companies install solar PV equipment on your roof, but they retain ownership of the equipment. You lease the equipment from them for a fixed monthly cost or a percentage of the system's production, which works nicely if you save more on your power bill than the monthly cost of the lease. Or alternatively, you can use these programs to lock in the cost of your energy so that when it rises, you won't be bitten like everyone else. You don't have to have any cash upfront, and this is often a very popular way to finance a big purchase, as evidenced by how many vehicles are leased on the same basic principle.

Ultimately, it's almost always a better investment to finance the equipment yourself and purchase it rather than lease it. But many people simply don't have the cash, they want to put their money into something else, or they don't want to sign up for a loan that appears on their credit report.

Be aware that, just as with cars, solar equipment has a *residual value* (the defined value that is estimated at the beginning of the lease; this amount may or may not be the actual market value) at the end of the lease. If your solar equipment has decreased in value for some reason (not super-likely, but it can happen), you may have to pay a big chunk of change to get out of the lease.

## Buying an energy-producing plot of land

Another concept that's likely to grow in the coming years is an ownership concept in which you don't actually install your solar equipment at your residence. Instead, a company buys land in Nevada or somewhere really sunny. The company divides the parcel up into small plots, each around 40' x 40'. Using cash or some other kind of financing package, you buy the plot of land

and a solar PV system that's installed on your plot and connected to the grid. The energy production belongs to you, and the equipment belongs to you. In practical terms, this is really no different than if the equipment were at your house, except nobody can see it, and you don't have to maintain it. Plus the equipment is probably in a sunnier location without any shade hindrances.

Whether or not this concept works in your area depends on whether a utility company will allow it. Most won't pay you for excess power averaged over the course of a year. When you install an intertie system at your house, if you generate more power than you use, you've wasted that power because you won't get anything for it. Obviously, with this remote location setup, there's no power use, only generation, so the entire thing earns credit. The good news is that the idea makes so much sense that governments are beginning to require utilities to accept the program, whether or not they want to. (Okay, they don't want to; why would any business agree to reduce its market?)

Remote ownership plans have plenty of advantages for the average homeowner. You can buy your solar equipment for much less. The company doesn't have to come to your residence to install the panels. All the systems are basically the same construction but in different sizes. Maintenance is straightforward and cheap, and the land is cheap. The productivity is high, and because the electric circuits are conjoined centrally, efficiency is much better. It's estimated that if 100 square miles of Nevada desert (of which there's no shortage) were covered with solar PV panels, the nation's electrical needs would be met with almost zero carbon emission.

# Working with Banks (Their Way or the Highway)

Banks have very rigid rules and procedures for lending money, and the people you'll be talking to have absolutely no power to alter these. When you're ready to talk to someone about getting cash for your project, you need to package your proposal to meet the bank's expectations, or you're wasting your time. Keep in mind that you, the borrower, aren't the bank's primary customer. A bank's customers are the people who give it money, not the people who borrow it. Banks make profits by lending you money at higher rates than it has to pay its customers for the money.

Banks are also concerned with risk, and they constantly strive to make loans at the lowest risk possible. When somebody defaults on a mortgage, especially a second mortgage, it's very costly for the bank.

If you want to do business with a bank, present yourself as a low-risk proposition. Get all your papers in perfect order. Get a copy of your credit report and make sure there are no surprises (banks hate surprises!). Look smart, be smart, and be prompt. When they demand more information, don't whine. Always remember that most of what loan officers do is search for a reason to deny you a loan; that's how they reduce the bank's risk. Basically, all they can do is say *yes* or *no.*

Don't take the process personally. It certainly isn't personal. You are but a spoke in the wheel or, worse, merely a Social Security number in the flesh.

Your property is the collateral for the loan, and it's critical in the risk-reduction process. To convince the bank that your property is ideal collateral, you have to convince the bank that if you default on the loan, it'll be able to sell your house quickly for a good price. Houses that are odd in character (off-grid home! *Beware!* — see Chapter 18) or that require modifications prior to sale (purple and yellow paint!) can cause hiccups. Banks want vanilla flavors for the same exact reason that you see mostly vanilla ice cream at the grocery store — because everybody wants it.

Keep in mind, however, that reducing risk isn't necessarily bad because you and the bank are on the same wavelength here. The bank wants your house to be worth as much as possible, and so do you. If you want to find out how to make your house more attractive as collateral, ask a real estate agent how to make your house worth more money.

# Part V
# Buying, Selling, and Building a Solar Home

The 5th Wave          By Rich Tennant

"They're solar panels. We're hoping to generate enough power to run our tanning beds."

# In this part . . .

In this part, you're talking serious money, with potentially fabulous rewards. The fact is, many solar technologies are relatively new and untested, and if you're planning on living in a solar home, whether you buy it or build it, you're going to have to make wise decisions because you simply can't rely on anybody else to make them for you.

I show you how to decide between buying a completed solar home and finding an existing home to convert to solar. I explain how to determine how much a solar home is worth versus a conventional home. I also show you how to find the right lot for a new solar home and how to design a good solar home to match that lot.

# Chapter 21

# Building a Solar Home from the Ground Up

. . . . . . . . . . . . . . . . . . . . . . . . . . . . . . . . . . . . . . . . . . . . . . . . . . . .

## In This Chapter

▶ Finding the right lot to build your solar home on

▶ Matching your perfect lot with your home design

▶ Considering the layout of your solar home

▶ Getting into the specifics of materials

▶ Biting the bullet and going for it

. . . . . . . . . . . . . . . . . . . . . . . . . . . . . . . . . . . . . . . . . . . . . . . . . . . .

Starting from scratch is the best possible way to get a solar home. By building a new home, you can control all the factors, including both passive and active. You can design the rooms exactly the way you want them. And you have good equity on the day you move in. When you build your own house, you also cut out all the profit margins that developers and financiers normally take. But building from scratch is a demanding process, full of surprises. To get an idea of the difficulties (as well as the rewards), read *Building Your Own Home For Dummies* (Wiley) by Kevin Daum, Janice Brewster, and Peter Economy.

You can usually build an energy-efficient solar home for the same cost as a conventional home. When you consider the fact that you won't have any energy bills, a green home actually costs less in terms of your monthly cash flow. This savings means you can afford a larger mortgage, which in turn means you can build a more expensive home (Note I did not say bigger!).

For the same amount of money, you can build a large, sparsely outfitted house, or a smaller house with more expensive fittings, appliances, and so on. The smaller the house, the better the potential for energy efficiency. So the first rule of thumb for an efficient house is to keep it small and appoint it well. This isn't really a problem if you pay attention to a few location features and design guidelines, which is what this chapter is all about.

# Following the Basic Rules of Home Building

Building a home is a major project, probably the biggest one you'll ever undertake. By keeping in mind a few basic rules, you can maintain your sanity and establish a better working relationship with contractors, designers, and others involved in the process.

The one overriding rule that you should never break is "Be patient." You're going to live in your new home a long time, particularly if you do a good job and end up loving it (why should you expect anything less?). To get everything just right, the whole project will probably take a couple of years. You'll make compromises, but impatience should never be the reason why you give in. Perhaps that perfect lot is a once-in-a-lifetime opportunity, even though it's not quite right for your solar home. But you can make design adjustments to account for certain compromises, so insisting on perfection is impractical and not really necessary.

Don't rely on your home being completed by a certain date. Some people who build a new home sell their current house long before their new place is ready, and they have no choice but to be impatient. Just be patient and have frequent conversations with your builder so you know when the construction process is behind or ahead of schedule.

The second rule is to take your new-home project seriously. Obviously, you want your new house to be the best it can be — but you'll probably be working with a general contractor, too. Keep in mind that building homes is his livelihood, and he probably has a family that counts on a steady income. As a result, contractors are very cautious about committing to a big job. They have all, unfortunately, seen potential jobs go away at the drop of a hat, and they've been left in the lurch. So never treat the process casually. Your decisions have major consequences for others, and you have a responsibility to proceed with respect for the ways in which you are influencing others' lives.

# Location, Location, Orientation

Everybody's heard the Realtor's favorite maxim, "Location, location, location." With solar power, this consideration becomes slightly modified, and orientation becomes a critical factor.

You're going to have to be a lot more fussy about your lot when you build a solar house. You want a piece of land that allows you to orient your house with a good southern exposure for PV panels and water heating collectors, and also for windows, sunrooms, and living spaces. On some lots, this is very difficult to achieve because of the street orientation and the location of trees and other shade potentials.

A barren lot is worth much less than a landscaped lot, but if the landscaping doesn't enhance the solar potential, or actually reduces it, it's useless, or maybe even a negative. Of course, you can raze everything, but that's at odds with the underlying green philosophy. On the other hand, you may have no choice, and then it will be incumbent on you to make things good with Mother Nature.

Of course, you need to keep in mind the features every homebuyer looks for when choosing a location (neighborhood, size of the home and lot relative to nearby houses, and so on) because one day you may need to sell your home, and potential buyers will consider these things. But for now, you need to find out two things:

✔ Whether you can build your solar-powered home in the neighborhood you're considering

✔ If the piece of property you're eyeing allows you to maximize the sun's energy

This section can help you answer these concerns.

## Evaluating local building codes and regulations

As you consider various locations for building a solar home, the first thing you need to answer is whether the local building codes support such a project. Unfortunately, some communities are behind the times and don't have laws and codes that are user-friendly.

You should start with the local building department (at the county headquarters) and discuss your plans for a solar home with a representative. During the meeting, you can find out what solar features are allowed and which ones are no-nos. You can also get a feel for how onerous and expensive the building process is going to be. Be sure to ask whether the department subsidizes solar homes and how you can get in on that action. Find out if other solar homes have been built in your area or the particular neighborhood you're considering.

## Out of sight, into court

I know of some homeowners who decided to put the solar panels at the side of their house so they couldn't see them, but their neighbor's could. This rather selfish attitude got them sued by the neighbor and the homeowners' association for violating the "visual appearance" clause. The association won because the judge found that the homeowner himself was declaring the panels visual blights because he didn't want to look at them either. Everybody lost a lot of money, even the winners. When it comes to rules, try to work within them, and never be stubborn about right and wrong.

If your conversation with the folks at the local building department goes well, your next step is to contact the homeowner's association where you want to build, or read through its charter. Here's where you may find the biggest contradictions. Many association bylaws ban solar panels because they're ugly; these restrictions are probably illegal. Nevertheless, you can try to comply with the rules against having solar panels in public view by orienting your house so the panels face south but aren't visible from the street. This requires the front of your home to face north and the back to face south.

If you still have questions about what you can and can't do, find a local contractor with experience building solar homes. He'll be able to answer a lot of your questions very accurately.

## Considering the solar potential

When you build a house, there's more to the story than just sun exposure. Solar is a general philosophy and takes into account water power, wind power, breezes, landscaping, thermal mass, and so on. Start a log book, and for each location you're considering, record facts, figures, and your observations for each of the following aspects. (Chapter 5 goes into detail on how to evaluate solar potential.)

### Know your solar exposures

What latitude are you in? What is the sun's path over the course of a year at that latitude? For each site, stand in the middle of the lot and plot the sun's course (you can make a graph, as per Chapter 5). You may have to do some guesswork about what the situation will be from a rooftop that hasn't been built yet. Different spots on a lot will have different solar exposures with different shading issues.

It goes without saying that you don't want to cut down trees. Look for deciduous trees and try to envision your house with those trees on the southern exposure, but not shading solar collectors on the roof. you can plant trees, but it will take a long time before they can rival natural, healthy, indigenous trees.

### Analyze prevailing winds

Natural breezes are absolutely free and can make a major difference in the comfort of a home. Look for locations where hills magnify breezes in the summer. In the winter, you want natural brush and landscaping to block the cold winds from the north. (Chapter 13 addresses ventilation and cooling issues.)

To thoroughly gauge the winds, you need to visit a potential location at different times of the year and in different weather patterns. Be patient in order to be thorough.

### Determine the water rights and drainage

Do you need a well? Wells have their advantages and disadvantages. You won't have a water bill, but you'll need expensive equipment that requires utility power. You can install a solar-powered well with a storage reservoir (see Chapter 14), and your water will be free forevermore. But wells can dry up, and the water can be of questionable quality.

You also need to determine the property's natural drainage. Some locations simply won't work for a home unless you radically alter the property's contour, which goes against the green mantra. How much water will your property require? Are you going to landscape? How will that affect drainage?

### Look in to other factors

If you can, camp out for a day or two on a prospective lot. At the very least, visit at midnight, in the morning, and in the afternoon. Visit at different times of the year, if that's an option. What sounds do you hear? Who is moving around and when?

And don't forget to determine what may happen in the future. Are you going to have a shopping mall next door five years hence? How about a freeway? Or an airport? To find out about these kinds of things, ask your Realtor and at your county building department.

# Designing Your Solar Home

Your goal in building a solar-powered house is to maximize the benefits of Mother Nature and minimize her drawbacks. So when you design your home, putting the rooms, walls, and windows in certain places and not others makes a huge difference. You should think about your house on a couple of different levels:

- ✔ How your home and lot can work together
- ✔ The overall size of the house
- ✔ Exterior factors that affect solar usage
- ✔ The location of the rooms inside the house

The following sections delve in to each of these items.

## Taking advantage of your lot's features

When you plan to build a solar house, you want to find the right lot first, and then get busy on the blueprints. You don't want to commit to a particular house design, and then insist on finding a lot that will bear it out. Each lot will support a different style of house with a layout that maximizes views, breezes, landscaping, and so on. So your house design should be a function of the lot.

You can take advantage of many of the existing features of your lot as you design your house. When you consider the terrain and landscaping, think about how those features change through the different seasons. How do the land, trees, and any other nearby structures influence the wind, precipitation, sun, and so on? Design on a 12-month basis, not just on the current season.

Hillsides work very well for providing insulation via earth berms and half basements. If you choose a hillside facing south, you can get two floors' worth of good sun exposure while enjoying northern insulation.

Deciduous trees work well on western and southern exposures. A house usually feels much nicer with direct sunshine in the morning, so try to keep the eastern shading to a minimum if that's your goal.

## Getting the basic principles right

Small is beautiful. A smaller house uses less building materials, is cheaper to maintain, requires less HVAC capacity, uses less energy, and so on. You can make a small house every bit as spacious as a large house if you

- ✔ Avoid long, wide hallways
- ✔ Combine utility functions like laundry and storage
- ✔ Put in less bathroom space (make the bathrooms tall, as opposed to wide, and you'll get a spacious effect)
- ✔ Forget both a living room and family room and combine them into one central great room

Your house will also feel much larger with a nice sunroom off the living area. And finally, with good, unimpeded views to the outdoors, a house feels much bigger than one with small windows leading nowhere. This could be the most important aspect of your entire design and is consistent with the green philosophy.

When you design a small house, create central living areas, which work much better than a meandering house with wings. Consider the ratio of square footage of living space (exclude utilities, hallways, and other areas where you spend virtually no time) to total surface area of the outside of the house. A house with a lot of wings is difficult to heat and cool, requires more maintenance, and simply does not need to be. It's a sign of surrender, not conquest.

## Starting on the design from the outside

Probably the most important factor for getting the most solar bang when building a house is to orient your house in the right direction on the property. The long axis of the house should run east to west for the greatest southern exposure. The front of the house can face south or north. Up to 30 degrees off true south will work well enough. If you've got a squared-shaped house, then you will want to put the family room and kitchen on the southern front to maximize your solar exposure.

The other thing I recommend is to base your design on heat movement; try to exploit the chimney effect and the greenhouse effect. Of course, you'll want to take advantage of natural breezes by venting appropriately (see Chapter 13).

### Up on the rooftop and what's underfoot

Roof exposure is critical for locating collectors. The roof is the ideal place for them because you'll have fewer shadows; plus, it's easier to hide the panels with a clever design. You don't need to locate all your solar panels on one expanse of roof, although visually it's usually more acceptable.

You can mount the collectors on the ground if you have enough room. But this requires wire runs and is more difficult to hide because hiding the collectors usually involves some sort of shading (which kinda defeats the purpose!).

Hard rectangular shapes aren't conducive to passive solar, so plan on breaking up the roof lines (which always looks better anyway). When you mount your solar collectors on the roof, the pitch should match the optimum solar potential angle. This is related to your latitude (see Chapter 5).

The more thermal mass, the more consistent the temperatures in the house. Slab foundations are best for this reason. Solar rooms with concrete floors and masonry walls work well, too.

### *Windows shouldn't be a pain*

The size of your windows affects the overall insulation properties of your house. Larger windows, of course, are less efficient, but they let in the most ambience as well, so it becomes a balancing act. You can achieve interior lighting by use of solar light tubes (see Chapter 9), so you don't necessarily need windows for light.

Reducing the number of windows on the northern exposure helps to increase insulation. If your best view is northern, the light will be cold and dull, and the windows on that side won't provide much solar heat, but will allow a lot of heat to escape. You may want a narrow house where the family room enjoys northern views while at the same time takes advantage of the southern exposure.

Good solar houses always have well-designed overhangs over the windows, porches, and doorways, particularly in the family room.

### *Strategically placing the garage*

Your garage can be located in the least desirable area of your lot. A western exposure can be hot and uncomfortable in the late afternoons and evenings, so you can put a garage on that side. You can also achieve a buffer from cold winter winds by placing the garage on the cold, windy side of the house. Or perhaps a view from one side of the house is unattractive, in which case the garage should go there.

## *What living space works best where*

Consider time of day versus room usage in your layout. Are you in a home office all day? Do you want to wake up in the sunshine? Think about the following points when you design your interior spaces:

- ✔ Family rooms generally work best on a southern exposure. Your family will spend most of their time there and get the most benefit from solar potential.

- ✔ Eastern kitchens are more congenial because you avoid the hot afternoon exposures while ensuring a bright morning wakeup call. A southeastern breakfast nook is ideal.

- ✔ A master bedroom on the northern front stays cooler, quieter, and darker.

- ✔ Put closets, bathrooms, laundry rooms, and other utility-type rooms on the north. You can tolerate poor lighting and cold exposures better in these rooms, so it won't matter if the windows are smaller.

- ✔ Sunrooms and windows are always best on the south where they receive the most sunlight. Use screened-in porches to shade windows on the east and west sides of a house.

Open airways make a house seem more spacious, as well as ensuring natural air movement. Avoid long, meandering hallways and rooms with odd angles and high, unventilated ceilings.

# The Devil's in the Details of Fabrication

You've defined the general design of your house. Now it's time to specify the particular materials and devices.

## Floors and windows

Tile floors have more thermal mass than wood, cost less, last longer, and take less maintenance. You can run a radiant heat system under your tile floor for even more comfort in the winter (see Chapter 12). You can set your thermostat down a few degrees and feel just as comfortable as in a conventionally heated house.

Avoid aluminum-framed windows because they conduct too much heat. Specify double-pane windows, at the very least. You can get triple-pane, as well as special windows with gas insulation between panes, but the cost goes up markedly. You're better off simply using fewer and smaller windows on the north and other fronts where there's no sunshine exposure.

Don't skimp on the quality of your windows because windows determine the value of your house to a large degree. Spend some extra money for windows with nice-looking frames and special features. At the very least, choose windows that are easy to operate because you're going to be opening and closing them a lot. If money is a concern, you can spend some extra dough for fancier features on the few windows that are important to the ventilation scheme.

Window blinds, awnings, and solar screens (see Chapter 9) provide visual appeal while retaining heat or preventing sunshine. After all you've been through, don't skimp here because these can make or break the ambience of a house. It's a common error in home building to forget about the stuff you'll need after you move in. Then it's not in the budget, which is probably going to be tight.

## Appliances and utilities

You've got a much better selection of energy-saving devices today than you had just a few years ago. Obviously, you want to install energy-efficient appliances, heating and cooling equipment, lighting, and other active devices.

You should use the smallest HVAC system possible for the amount of square footage in your home. And install ceiling fans in the rooms where you spend a lot of time. They save money and look good in the way they break up a ceiling (see Chapter 13).

A stove, whether gas, wood, or something else, is always a good idea in the family room because you can use it exclusively 90 percent of the time for your heating needs. Most of the time you don't need to heat your entire house. In the same vein, room heaters and air conditioners are much cheaper to operate than a whole house HVAC system.

To conserve water, use as few sinks as possible and keep the piping runs as short as possible. Showers are more energy efficient than bathtubs, although it won't matter if you have a good solar water heater system and use it properly.

Lighting without electrical is always desirable. Use solar tubes, reflectors, orientations, and so on. Celestories can provide both heating and lighting over the course of the year. Install PV panels for whatever lights and other devices that need electricity to operate.

If you're building a swimming pool, consider a solar-powered pump. It will be expensive up front, but you'll never pay a cent to power it. This is especially true if your house is in a region with TOU metering (see Chapter 11 for pool systems).

## Roofing and ventilation

Roofing shingles should be reflective and insulative, unless you're more interested in a hot attic in the winter than a cool one in the summer. In either case, consider how you're going to vent your attic in both seasons.

Consider placing radiant barriers in the attic space. Put these in while you're building the home because they're much more difficult to install after the fact. And the material is relatively inexpensive, considering the potential benefit.

To keep your house cool in the summer without running the air conditioning, install an attic vent fan, controlled by a thermostat switch, or a whole house fan located in a central spot (see Chapter 13).

As long as I'm talking about heating and cooling, I'll remind you that insulation should be as thick as you can afford.

# Time to Start Pounding Nails

Hiring a contractor and going through the construction process are pretty much the same for solar homes as they are for conventional homes. You don't need a contractor who has built solar homes before; the passive aspects of the solar design are part of the blueprints, and any contractor should be able to do the job properly.

As far as installing the solar equipment itself, a good plumber should be able to install the solar water heating equipment. If the plumber you've contracted with can't do it, ask him to subcontract the installation to a solar installer. The entire solar water heating system should be installed as part of the finished house.

The PV system is another story, however. The problem is that you just don't know how much capacity the system will need. You won't know how much energy your house requires until you've lived in it for a year or so. The energy requirements depend on your personal habits and how the house is affected by weather patterns. If your system is intertie, which it definitely should be if you have the option, you won't get anything from the utility company if you provide it with the extra power that you don't use (see Chapter 17 for more details). Therefore, the best bet is to wait a year or so before you install your PV system.

The problem with waiting to install the PV system is that you probably want to finance it as part of your new house. If you wait a year and decide to put in a $25,000 system, you'll need to come up with the cash then. You may be able to get a home equity loan, but the terms won't be as favorable as if you wrapped the PV financing into the original home loan. So if you want to take care of all the financing at once, estimate your energy needs, and install a system that's smaller than you think you'll need. You can buy an oversized inverter for the smaller system so that when you're ready to expand it, you only need to put in a few more solar panels to zero out your power bill. And you may find that the smaller system keeps your power bill in the lower tiers, so you don't have much of an economic incentive to install that extra capacity (see Chapter 17 for more details).

When you design the house and start construction, make the roof pitch and construction optimum for PV panels. And design the PV system layout so the installation will be quick and easy when the time finally comes. Leave room near the fuse box for an inverter and switches.

# Chapter 22

# Buying and Selling a Solar Home

. . . . . . . . . . . . . . . . . . . . . . . . . . . . . . . . . . . . . . . . . . . . . . . . . .

## In This Chapter

▶ Looking for a conventional home to turn in to solar

▶ Inspecting and buying existing solar homes

▶ Selling your solar home

. . . . . . . . . . . . . . . . . . . . . . . . . . . . . . . . . . . . . . . . . . . . . . . . . .

*T*he bulk of this book is dedicated to solar projects you can do in or around your home. But perhaps you want to move in to an existing solar home where everything is already in place. You won't have to do any projects at all — they're already done for you. Or perhaps you want to buy a conventional home that is more suitable for solar systems than your current home, and you plan on installing the solar systems in your new place.

In addition to the usual pest, roof, and other inspections normally done for a real estate transaction, you will need to have energy audits and solar potential inspections. In particular, you need to understand solar equipment because you are going to have extra responsibilities with a solar house, and those responsibilities lie squarely on your shoulders. Add the facts that solar is relatively new and most real estate agents don't thoroughly understand the ins and outs, and you realize how important it is for you to be wise about what you're getting into. This chapter helps you make the most of buying (and later, selling) a solar home.

Whether buying or selling, your goal is always to get the best deal you can. With conventional homes, this is a familiar process for everyone involved. But it gets a little trickier with a solar home because the market is relatively untested and things are changing fast. You need to understand how solar affects home prices in your area. A solar home is almost always going to cost more than a conventional home, and whether you're buying or selling, you need to understand how to put a number on this difference. This chapter shows you how.

# Buying an Existing Home with the Intention of Installing Solar

On a drive through almost any neighborhood, you generally see one, two, or more homes for sale. The vast majority of them won't be solar, but who cares? You can install solar equipment on any home. The majority of this book addresses how to do this.

The upside to buying an existing home and converting it to solar is that the options are much greater. You have more control over the house's location, its layout, and its price and size. Not to mention, you have more houses to choose from. In some areas, solar homes are a real rarity, and you may not even be able to find one that's for sale. Restricting yourself to buying an existing solar home will probably mean you'll make some big compromises on the type and style of home you want. This is a questionable strategy, given how important a home is in your life.

Even if there are solar homes in your area, they probably won't be for sale. Homeowners rarely install expensive solar equipment right before they sell. It takes too much work and financing to be worth it, and the returns — while good on paper — are risky. If you do find a solar home for sale, the solar equipment is probably older and not as efficient as new systems that take advantage of the impressive technological improvements realized the last decade or so.

Of course, buying a house and making it solar has other benefits and drawbacks. I cover these in the next sections.

## Dwelling on the pros

For a do-it-yourselfer, what's better than having a blank canvas on which to install brand-new solar projects? That's what you get when you buy an existing, un-solarized home. Solar projects are fun and interesting — much more so than moving walls and redoing plumbing. Plus, you get the satisfaction of seeing tangible results on your power bills.

When you install your own equipment, it will be brand-new with a full warranty. New equipment is always better than old because technology is always improving. And when you put a system in yourself, you'll know the equipment better, and you won't have to trust a seller's claims of performance.

You can also get cost savings when you do the work yourself. As a do-it-yourselfer, you can install equipment for much less than a professional solar contractor. So the value you place on solar equipment will be less than what a seller is probably going to demand for a completed solar home. Plus, you may be able to get subsidies or tax incentives to install new solar equipment.

After you've lived in the home a while and know your energy usage, you can install equipment as the need arises. You may want to install a small PV system with the expectation of future expansion when utility rates increase. Also, net metering (see Chapter 17) may not be available right now, but it's slated to be available in the future.

In some cases, you may be able to demand that some improvements be done before you buy an existing home. Perhaps you can get a seller to install some solar equipment or some energy-efficiency equipment as a condition of sale. If a water heater is shot, perhaps you can demand a solar water-heating system be put in its place. If the windows have seen better days, maybe you can talk the sellers into installing some good double-pane windows. If you get a motivated seller, you can leverage your position. You may also ask for money (to be paid at escrow) for solar improvements and in this way tie the financing into your original first mortgage.

## Considering the cons

Buying an existing, non-solar home that you plan to convert involves lots of unknowns: How much power will you need? How much money will you sink into solar systems and where will it come from? How much time and energy will converting the house require?

With an existing home, you'll likely need to get an energy audit, and perhaps live in the home a few months before you know how much power you need. When you buy a completed solar home, there's a track record of performance, and so the risk on return is less.

You'll probably finance the home with a mortgage, which won't include money for your future solar improvements (although there are financing arrangements that you can find that will finance future improvements). You may be able to get a home equity loan to pay for the upgrades, but ultimately, this will cost you more than if you simply buy a completed solar home and wrap all the extra value into the primary mortgage.

After you've figured out where the money is going to come from, you need to determine who's going to do these projects and when. It takes work and time to do projects, and you may have to deal with a number of contractors. Many people have experienced contractors often enough to know that they want to avoid them in the future. If you've never worked with a contractor, be aware that it can often be frustrating. Talk to friends, neighbors, and coworkers to get a sense of local contractors' quality of work.

If you decide to take on the solar projects yourself, consider your resources. Are there good hardware stores nearby? Where will you get your supplies? Do you have enough time to tackle the projects? Do you know what you're doing? (I'm sure you do after reading this book. Just don't ask your wife for her opinion unless you're itchin' for an argument.) You may be able to get a guarantee of performance when you buy a completed solar home. But when you do the installs yourself, you don't get any kind of guarantees, or at the very least, they are hazy.

## Getting the information you need

If you're going to buy an existing conventional home and install your own solar, you need to carefully and accurately evaluate the home's suitability for solar. Here's a checklist:

- ✔ **Get a professional energy audit, if possible.** The seller may let you do one, but it's a tough request to grant before you've made an offer on her house. It's one thing to walk through a house; it's quite another to climb around in basements and attics poking tools into things. Would you want somebody doing that to your house before you knew if he was serious? Inspections are normally done after an offer is made and accepted.

- ✔ **Evaluate existing energy equipment for potential improvements.** Is the HVAC old? Perhaps you can demand a new one be installed. How about wood stoves or gas stoves? Are they working properly, or do they need to be replaced? If they're not in the house, can they be installed? Should they be installed? Attic vents? First of all, do they exist? Secondly, do they make sense. The latter question is tough to answer; ask the energy auditor to take a look and make some recommendations.

- ✔ **Make a list of all the solar investments that will work in your new home.** Then estimate the costs for doing each project, based on the construction and layout of the house. This is relatively easy to plan for active solar, like PV and hot water. But it's going to be tougher to decide what will work for passive, and you'll probably find that moving walls and cutting into ceilings isn't really worth the cost, given the potential savings on your power bills. Passive solar is much easier to design and build into a home than it is to modify an existing home for.

- ✔ **Evaluate the outside of the house as well.** Do the trees shade the way they should? Are they deciduous? If not, what will it take to rectify the situation? Can you add a sun porch, or a trellis, that will change the nature of the layout? How about adding overhangs over windows (see Chapter 8)?

- ✔ **Read Chapter 21 on building a solar home from scratch.** It details all the things you want to do when you design an ideal, new solar home. When you're looking at a prospective home, compare its features with the ideals expressed in Chapter 21. And who knows; you may become convinced that your best bet is to build your own custom solar home.

# Buying an Existing Solar Home

Unfortunately, not too many solar homes have been built yet. This is going to be changing quite a bit as solar power catches on, but the reality is that if you want a solar home, you're going to have to look long and hard, and you may be disappointed with what you find. You can build your own solar home (see Chapter 21), but that's a long, difficult undertaking.

Solar communities are springing up around the country. All the homes feature both passive and active solar, and the designs have been done by professionals who know how to make solar systems work to their fullest potential. In some of these communities, the houses each have their own solar PV and hot water panels. In other communities, collective solar facilities are shared and maintained by a central administrative function. That way, individual homeowners don't need to worry about the hassle and expense of operating and maintaining themselves.

The trend toward solar communities makes a lot of sense, and you should pursue finding a house in one of these neighborhoods, if possible. All your neighbors will have the same equipment you'll be using, and you'll have a community of support for technical problems and user issues. A ready army of professionals will be available to help when needed, because the entire community has an interest in maintaining trustworthy contacts.

But for most buyers, it's going to be an individual enterprise. As usual, "Let the buyer beware" is the rule. If you find a solar home that you're interested in, you need to do the usual inspections, but you also have to make sure that the solar equipment is sound and does what it's supposed to do. Keep in mind that solar equipment increases the maintenance responsibility of any house; you need to know how to use the equipment and how to take care of it, even if it's all in perfect working order.

## Checking out the home

Armed with all of the smarts you've picked up reading this book, you should be able to walk through a solar home and generally figure out if it's up to snuff. If you see too many things that look outdated or that don't seem to be working properly, move on. You want to take advantage of as much of the sun's energy as you can. This section explains what you should look for as you evaluate whether a particular solar home will meet your needs.

An energy-efficiency audit is always a good idea, not just for the solar equipment, but for the entire house. You can have this done as part of the usual due diligence, or you may want to pay for it upfront before you start getting serious about the house. This audit will also reinforce your conclusions from your evaluation of the home.

### Poking around outside

Before you even set foot in the house, you can get a feel for how effective the solar features are by looking at the home's orientation, structure, and landscaping. (For information on the ideal solar home, see Chapter 21.)

- ✔ Which way does the home face? The best bet, for solar, is a southern front, but it doesn't mean that the home needs to face the south. A good solar home will have southern windows, and the family room will be on the south, preferably with large windows shaded by deciduous trees.

- ✔ Are the deciduous trees in the right spots?

- ✔ Are there solar tubes showing on the roof? Skylights?

- ✔ Do the windows have overhangs?

- ✔ Are thermal masses being used effectively?

- ✔ Look at the roof and see if the collectors are aligned optimally. Are they facing east or west? Could you make a few changes and get more out of the equipment? Are there trees shading the collectors that you could cut back to get more productivity? Are the collectors dirty or damaged?

You'll get a roof inspection as a matter of routine. Keep in mind that not only will you need a new roof at some point, but if solar panels have already been installed, you'll also have to pay to get the equipment dismantled first, and then reassembled after the new roof is in place. You'll also be getting no solar production during this period.

### Looking at the home's design

You can change some things about a home's interior. Other things you have to live with. Here are some features to consider:

- ✔ Is the family room on the south, with big windows?

- ✔ Are the blinds strictly decorative, or do they also have functionality?

- ✔ Is the fireplace centrally located?

- ✔ An efficient home uses a large, central living space. Homes that are spread out with meandering hallways require a lot more energy to heat and cool, plus they use a lot more building materials.

- ✔ Utility spaces should be located in easy to access points.

- ✔ The kitchen should be adjoining the garage.

- ✔ Windows should be functional as well as decorative. The best designs have windows that are both functional and decorative at the same time.

### *Inspecting the existing solar systems*

You'll want to carefully inspect any existing solar systems. The best bet is to have a professional check them out. (Perhaps the energy auditor is qualified to do so; otherwise, call around, but be aware that a solar contractor may try to sell you a new system.) You can check the systems for yourself, which is a good idea, because if you buy the house, they're going to become your property, and your responsibility, so you'll have to know how to operate and maintain them.

✔ Find out whether the systems can be expanded. For example, is the inverter larger than its current output power? If so, PV panels can be added. A hot water system can usually be expanded by adding more collectors.

✔ Check the PV system. You can do this by looking at the power output on a sunny day near solar noon.

    **1. Check the panels and see how far off the optimum angle they are.**

    **2. Simply read the digital meter on the face of the inverter.**

    This will be the maximum amount of power you can expect to get out of the system. Does this comport with the system size? You can find out how many panels there are and what type, and use PV Watts (see Chapter 17) to determine how much output the system should be giving. If it's less than this, something is wrong, or the system is old.

With this information, you can determine the maximum system output, and from this amount, you can estimate the annual generation, that will yield your cost savings. (Chapters 16 and 17 explain PV systems in great deal.)

✔ Are there local building department permits for the equipment? Why not? You may want to insist on the seller getting this done, although she's probably going to balk because it's a big hassle.

### *Asking about installation and upkeep*

In all likelihood, the seller of the solar structure (how's that for alliteration!) will be eager to talk to you about his home. Take advantage of his passion for his home and lifestyle and find out the following information:

✔ When was the equipment installed?

✔ Who installed the equipment?

    If it's a do-it-yourself job, have the work checked out by a pro, or check it yourself, especially if you're also a do-it-yourselfer.

✔ Where was the solar equipment purchased? Is the company still in business? Why not?

✔ Who designed the solar equipment? If it's passive, was the architect or designer qualified? Temper this answer with some common sense. A pro can design a poor house, and a novice can design a great house (especially if she's read this book!). Ultimately, it's performance that counts, not pedigree.

✔ How much warranty is left? Is the warranty transferable (this is important)? Have any warranty repairs been done? If so, pay attention to what went wrong. If several warranty repairs have been done, beware, especially if the warranty is about to expire.

✔ What records have been kept regarding system performance and maintenance?

✔ Can you look at the manuals that came with the equipment? You may be able to get these on the Internet, but a conscientious owner (the kind you want) should keep them. Read through the manuals and you'll understand the equipment and what it's going to need by way of maintenance and repair.

If you buy the house, ask the owners to leave behind the manuals and any maintenance records or other relevant paperwork for your reference.

### Sleuthing out energy costs

If everything else about the home checks out, take a look at the power bills. If the seller doesn't have them, she (and sometimes you) can get them from the utility company. If you can't get any power bills, something is wrong. The seller should be proud to show them off, as the vast majority of solar home owners are.

Of course, personal habits enter into the equation, so you need to temper what you find with a consideration of how the seller's energy usage may differ from your own. This may not be easy, particularly if you're not able to talk directly to the seller. How many people live in the house? The kind of cars they drive often signals how they use energy in their home. If they're all driving enormous SUVs, it's likely you'll be able to do better on energy conservation than they do. Or you can simply ask about their energy habits, although this is a tough nut to crack.

If some monthly bills show a total that blips up inexplicably, it could be because the solar equipment was broken at that time. Ask why the bill blipped, and take warning if the answer doesn't jibe with the seller's claims of how often the equipment went down.

You also need to consider *utilization,* which is how much solar equipment is being used versus its maximum capacity. For example, if the house has a solar water heater, how much of its capacity is being used to offset the current power bill? If it can put out twice as much energy as it is currently, is this important to you? If your family is bigger, you'll use more hot water, and you'll probably use more energy in general. This may or may not be reflected in higher energy bills.

You should also talk to the utility company and find out the following:

- ✔ Is there a net metering agreement. If so, does it carry over?

- ✔ What is the rate structure going to be? If it's not going to be the same as the one in place the last few years, your utility bills are going to be different.

- ✔ What are the new bills going to be? Now you've got some math to do, but luckily you have my expert advice to rely on.

- ✔ Can you change the rate structure if you want?

## Determining a solar home's value

A solar home is going to cost more than a comparable conventional home. To figure out a fair offering price, you can ask your real estate agent for an estimate on how much a conventional home is worth in your potential home's neighborhood, and then add the value of the solar equipment.

Of course, a major transaction like buying a house is never as easy as it sounds. You can place two sorts of values on a house: financial, based on hard facts (this is the one the bank is interested in), and emotional, based on how you feel about the house and what it's worth beyond monetary value to you. Combine these two values, and you've got a good idea of what you should offer for the house.

Some banks may not add any value for solar equipment (actually, it's more up to the appraiser, but that's a technicality). This can be a problem if you're trying to get a very high *loan-to-value ratio,* which is the amount of the mortgage you're applying for, divided by the sales price of the house. If you have a good down payment, it won't matter if the bank doesn't take the solar features into account. If the bank's disregard for solar becomes a problem, shop around for different banks. While some are completely ignorant of solar, others welcome it (see Chapter 20).

### Delving in to the data

When you're trying to figure out how much a solar home is worth, start with getting what are called *comps* in the real estate business. These are the recent sales prices of comparable homes with solar power in the same local area.

After you've determined the market value, you can look at the home's features and decide how those will influence the price. Here's a rundown:

- ✔ How much would it cost to have a contractor come in and install all the solar equipment that is already in place? A solar home's equipment should never be worth more than this, although buzz can sometimes send values higher. Plus, it's more difficult to do projects yourself, and a lot of people simply don't want to, so they'll pay more for it already in place.

- ✔ How much maintenance is there going to be? Who is going to do it? How do you place a value that? Do you have tools and expertise?

- ✔ What is your seller claiming is the increase in value due to the equipment? Is this reasonable? Are the promises reasonable?

Make sure to get in writing exactly what equipment is staying or going. In most states, real estate law clearly states that if a piece of equipment is attached to a house, it's part of the house. Doors are part of a house, whereas refrigerators aren't. In theory, all the solar equipment is part of the house. But this is often misunderstood, particularly with regard to swimming pool systems, which sometimes end up going with the seller to her new house.

Because you're buying a solar house, you have one additional piece to consider when it comes to determining financial value: the savings you expect to realize from the solar power.

- ✔ How much less are the energy bills than those of a comparable conventional home? Realize that you can buy more house when you don't have a monthly utility bill.

- ✔ Where do you think power rates are going to go in the future? If you think they're going to go through the roof, a solar home is worth more.

- ✔ How much is it worth to you to mitigate pollution? See Chapter 6 for a method of calculating this rather intangible issue.

- ✔ Are there tax incentives to buying an existing solar home? You may be eligible to get tax credits, retroactive rebates, and so on. See Chapter 20.

- ✔ What will owning a solar home do to your insurance rates? Call your insurance agent and ask how solar equipment might affect your rates.

### Listening to your heart

Many people wind up buying a home based more on emotions than on facts. In some cases, your heart (or your gut) is right; in other instances, you need to pay attention to all the red flags your head is throwing at you. Consider:

- ✔ Would you buy this house if it were exactly the same structure, but without any solar equipment? How much would you pay for it then?

✔ Would you want to live in this house if it weren't solar? Why not? How much do you value your compromise in comfort and aesthetics?

Now you've got a decision to make. How much do you want the house? That's the bottom line. Put your head together with your real estate agent's to come up with a competitive offer.

# Selling Your Solar Home for Big Bucks

The day will come when you need to sell your solar home. You want to get the most money you can for your home, of course. How do you accomplish this?

Some factors come into play long before you think about parting ways with your abode. As you install and maintain equipment, keep good records of everything. If you can't validate your solar claims, your systems may be worthless. Or even more detrimental to your cause, you may cast yourself as untrustworthy in general, because why would you be making claims if you can't back them up?

When you're ready to let the world know your home is for sale, find a real estate agent who knows solar houses and can help you capitalize on the "buzz" of solar homes. The right agent will let the right buyers know you've got a solar home for sale. At present, most people have no clear idea what solar means. They don't really care about a solar house, and these people aren't likely to pay more for your solar home than they would pay for a conventional home. As energy rates rise, this will change out of necessity. But one thing is always true of marketing: You need to reach the right market to get the best price.

You are going to price your home higher than a conventional home, and rightly so. But if you don't reach the right market, your home won't sell, and then you'll have a reputation in local real estate circles for trying to sell a home that's overpriced.

The initial entry of a home onto a market is the most important timeframe because the longer a house sits on the market waiting for a buyer to come along, the less interest real estate agents have in showing it. Even if you subsequently lower your price, they still aren't going to get as enthused as they do with a house that's new on the market and priced correctly.

It's very important to price a home properly upfront. The starting point is *comps* (comparable house sales prices in your area), but after that, it's a question of experience and an ability to read the market. A good Realtor can tell you exactly what your house will sell for. Most sellers, however, pressure the Realtor to list the house at a higher price because they want more (of course), but they're also unrealistic. If you hire a Realtor who is listed as a top producer, you'll get one you can and should trust. Top producers cost the same as other Realtors, but come with an important pedigree.

There is no formula for what you can get for a solar home over a conventional home. Home values vary greatly from region to region, but the cost of solar is relatively constant, so the percentage difference for a solar home is going to vary from region to region. The key is to get a Realtor who is on the same page as you and ask about her solar experience. Most listings appear on the Internet these days. Make sure to include the words green and solar as keywords. In the MLS listing, use the same words. Stress the fact that the utility bills are lower; you may even want to parade that fact on the top of any brochures and fliers that the Realtor produces. You can even notify local environmental groups, such as the Sierra Club.

Finally, make sure to read the earlier sections in this chapter on buying a solar home. Understand that now you're on the opposite side of the table. Throughout the process of showing the house, landing a potential buyer, and negotiating a deal, you and the buyer will be at odds over some things. They want to pay the lowest price possible; you want to receive as much money as you can from the deal. They want documentation and maybe some upgrades or repairs. You want to minimize your headaches and any additional expenditures. But on the other hand, you'll both want some of the same things — a smooth transaction and a house that lives up to its solar potential. With a little patience and some help from your real estate agent, you can make it all work out.

# Part VI
# The Part of Tens

"Solar conducting razor wire? I like it!"

# In this part . . .

**E**very handy, information-packed *For Dummies* book
ends with quick reference lists to help you along in
the process. I've packed good information into a few
words, all designed to elicit serious ideas and reflection
on your part as well as give you some direction on where
to go — and where *not* to go. And because this is the last
part of the book, I've allowed myself to let loose with the
humor that I normally keep restrained somewhat below
the surface — because all things solar should end on a fun
note.

# Chapter 23

# Ten Best Solar Investments

*In This Chapter*

▶ Identifying the best money-saving propositions in solar power

▶ Making your house look a lot better

▶ Creating a more comfortable home environment

*I*n this book, I describe hordes of projects and improvements you can tackle. Your personal situation dictates the type of solar investment that's best for your own home, but some projects stand out because they're practical for nearly every homeowner, in both monetary and aesthetic terms. In this chapter, I give you my list of the ten best solar investments, based on my own experience and the feedback from others who have also worked with solar.

# *Nurturing Mother Nature with Landscaping*

Planting hearty, healthy, happy deciduous trees in the right location around your house gives you cooler summers and warmer winters, but most of all, you can look out your windows and be reminded of why we care so much about our planet Earth.

Planting bushes, shrubs, and trees as windbreaks allows you to enjoy natural breezes in your home, without the sound of whirring fan blades to remind you of technology.

Plants create oxygen out of carbon dioxide, the modern bugaboo of environmentalism. If the world had enough trees, global warming wouldn't be such an imposing issue.

And finally, you get yourself outdoors and you get exercise when you do your own landscaping. There's a certain simplicity to digging a hole. It's about as close to nature as you can possibly get, and that in itself is justification.

# Installing PV Systems to Offset the Most Carbon Pollution

When you install a large PV solar system, you cut out a tremendous amount of pollution because our electrical power grids are extremely inefficient. For each kWh of energy you create with a PV system, you save three or four times that much utility-generated power, most of which comes from coal-fired plants in North America. PV systems allow for tremendous environmental leverage, and that will never change.

Strictly from a monetary standpoint, PV systems are becoming more competitive, and as energy rates rise, they will become better investments. But even if you pencil out the investment and it doesn't look too good (because your power bills aren't that high, or you don't get all that much sunshine), you can use a technique that I present in Chapter 6 to modify your financial analysis with a valuation of pollution mitigation, and you'll probably come to the conclusion that a large PV system makes sense.

In addition, when you install a full-scale PV system, you lock in your energy rates for a long time, namely at zero. If you think energy rates are going to rise precipitously, PV is almost always a great investment. And really — what else is going to force America off its oil addiction than high energy rates?

# Using a Solar Swimming Pool Heater

When you install a solar heating system on your swimming pool, you can use your pool more than you would without a heater. The water temperature is much more comfortable, and you can swim over a longer season.

The average swimming pool costs upwards of $25,000 to install. It takes up a big chunk of your backyard and requires a lot of maintenance. With all the time and money you put into your pool, it's probably the most expensive luxury you will ever purchase.

You can use a pool without a solar heater for about four months, but at the beginning and end of the season, the water is freezing, and you go swimming because you spent so much on the pool and you're bound and determined to use it, not because you're having fun. But when you install a solar heater, you can use your pool for half the year, and the water temperature is more comfortable for that entire time. A $5,000 pool system makes your $25,000 pool investment much more sensible. (See Chapter 11 for more on capturing solar power to heat your pool.)

From a pollution standpoint, a solar heater is ideal (zero pollution) compared to the sinful alternatives (propane and electric).

# Putting a Cover on Your Swimming Pool

I can hear the howls of indignation over this one. Swimming pool covers are a hassle. But for the price, they have a huge effect on how you enjoy your pool. Swimming pool covers basically accomplish the same thing as solar pool heaters (see the previous section) but cost about 3 percent as much (they're about $150). They keep a pool cleaner, so you save on chemicals as well.

If you're in a climate where you don't need a pool heater during the summer, you can extend the swimming season a month or two by using a cover only a month before and a month after you would normally use the pool. Chapter 11 explains additional benefits of swimming pool covers.

You can attach the cover to a retracting mechanism so that you won't hate your cover like you would if you had to fold it by hand every time you swim (and unfold it back onto the pool when you're finished). The cheapest manual retractors cost around $300. You also can install an electric retractor, but now you're talking about the kind of money you'd spend on a solar water heater, so you might as well put in one of those instead.

# Harnessing the Sun to Heat Your Water

Solar water heaters, when properly designed and installed, are great investments because they are much cheaper than PV systems, which are often out of the average person's financial range. Most homeowners can afford a few thousand dollars for a solar water heater system without taking out loans.

A benefit of solar water heaters is that you won't need to conserve on hot water, regardless of how high energy rates go. It makes no sense to use a solar system at less than its full capacity, because it saves nothing.

From a pollution standpoint, heating water typically comprises around 18 percent of your power bill. A solar hot water heater can offset around 75 percent of this, and you'll save equally on your carbon footprint.

# Lighting Your Yard All Night Long

For little cost, you can put a range of fun and interesting lights around your yard. They charge during the daylight hours and come on at night. They need little sunlight, given the amount of light they put out.

The alternatives are awkward, clumsy, and demanding. Utility powered, low-voltage systems have thick-gauge wires that you need to run around your yard, tripping people and getting chewed up by the dog. And the lengths of runs you can get away with are limited because the wire is so expensive.

Putting in solar lighting is as simple as one, two, three. And if you don't like the way things look, changing the layout is as simple as four, five, six. Try both static lights and the changing-color decorative lights. You can get a whole range of different mounting schemes, so you can place the lights anywhere. (Chapter 8 has more info on solar lighting.)

My experience is that the lights don't even need to be in direct sunlight. Put them under a tree, and they'll work. You also can get ones that have the PV panels connected to the light by a wire, so you can put the PV panel in direct sunlight and the light under your porch roof.

# Redecorating for Functionality and Appearance

Window blinds make a big difference in the overall look and comfort level of your home. Windows attract a lot of attention (they break up walls, which are monotonous), and they're a source of natural sunlight. The eye is naturally drawn to a window, particularly a big one in your family room or living room.

The fact is, windows are a major source of heat transfer. In the summer, windows let in too much heat energy by both radiative sunlight and conductive heat movement. In the winter, windows allow a lot of heat to escape by conduction. Your house would be much more energy efficient if it didn't have any windows at all, but this is absurd. The solution? Put in window blinds that have good insulation and reflective properties.

With the right blinds, you can significantly reduce heat transfer as well as reflect most incident sunlight. The functional effect is dramatic. The aesthetic is even more so, if you choose the right ones. (Chapter 9 has suggestions on what sorts of blinds to hang where.) Large windows in family rooms and living rooms are the best candidates for installing blinds. You get the most bang for the buck when you cover these large areas of glass.

# Putting Up Overhangs to Make Your Home More Comfortable

By using overhangs to shade your southern windows appropriately, you can increase natural warming in the winter and prevent overheating in the summer. You also can improve the natural light in your home by increasing the amount of sunshine you let in during the winter, when you want as much light as possible (it makes you feel warmer, and perception is half the game), and by decreasing it in the summer, when a lot of light makes you feel hot.

Controlling the sun as it shines into your house lets you regulate the temperature variations. Nobody likes a home where the temperature swings wildly over the course of a day. And temperature variations tend to make materials swell and shrink, which causes cracking and premature wear. When you install a well-designed overhang over a porch or sunroom, you minimize temperature variations.

Overhangs are very reasonable do-it-yourself projects (Chapter 8 has more details). You don't need electrical experience or plumbing know-how. There are usually no extraordinary weight requirements that entail consulting a professional engineer. And if you keep things modest, you don't need to get a county building permit or permission from an association design committee.

# Increasing Your Living Space

You can add a solar room onto your home for far less cost than a conventional room. You can put in nearly any size you want, and do-it-yourself kits are straightforward and well designed. You can build a solar room out of aluminum or wood, and you can put in however many and whatever size windows you want. You can incorporate a concrete floor (for maximum thermal mass), or you can use an existing wooden or synthetic deck.

If you do it right, you can build a solar room without getting a building permit (forego electric power and don't connect it permanently to your house). You can build a solar room out of plastic corrugated materials that cost very little. You can grow plants in a greenhouse year round, or if you want to get really exotic you can grow fruits, flowers, or vegetables. If you choose to build a sunroom, you can use it as extra living space when it's not too hot. If you insulate it well, you'll have a family room for about one-tenth the cost of adding to your house.

# Banishing Hot Air with a Solar Attic Vent Fan

If you've ever gone into your attic on a hot summer day, you know what real heat is all about. It can get so hot that it's dangerous. Temperatures over 160°F are not uncommon. All that heat stays up there at night, and it sinks into your house through the insulation in your ceiling. Most homes have passive, natural venting schemes in the attic, but that doesn't do much to get the hot air out of your house.

A properly designed solar attic vent fan can move a lot of air (see Chapter 13) over the course of a day. The system works hardest when there's a lot of sunshine. You don't need to run expensive electrical power up to the fan, which means you can install one just about anywhere you want.

As a do-it-yourself solar project, it's ideal because you get to use some PV panels, which is fun (if your definition of fun is cool hardware). You won't get any electrical shocks from the low voltages, and the tools required are minimal. You can complete the project in one day, and you'll discover a lot about your house by studying the layout and functionality of your attic.

# Chapter 24

# Ten or So Best Do-It-Yourself Projects

*In This Chapter*

▶ Mastering easy projects that you can complete in a day or so

▶ Climbing, screwing, pounding, measuring, charging, cooking

▶ Tackling challenging projects that will impress your friends

*E*ver since I was a boy, I have loved tools and the damage they can do. I have learned over and over that nothing ever works the way it's supposed to. I have come to accept the constant companionship of Murphy, who, I have also learned, is an eternal optimist. Nonetheless, I have forged on, tackling bigger and bigger projects, and making bigger and bigger messes of things. Ultimately, if I'm patient and keep a ready supply of bandages handy, I can get things to work somewhat satisfactorily. I love doing projects, especially ones I can design myself. If you're like me, this chapter is for you.

# Using Just About Every Tool in the Box to Install a Solar Water Heater

If you want to install a solar water heater, you have to be good at plumbing, electrical wiring, solder joints, PV panel installation, copper tube cutting, climbing around on roofs, figuring out what to do when things don't fit right, and that's just for starters. You need to understand water pressure and how it can cause pipe weld joints to burst at the worst possible times. You need to understand what happens in cold weather to water that might freeze in an exposed pipe. You need to know how to evacuate closed systems of air and how to burp and drain pipes completely, or else you'll be sorry.

You need to know how to choose the best kit among a horde of suppliers who have very poor literature and don't like to answer the phone very often. When they finally pick up, they tend to sound like you're an idiot for asking perfectly logical questions.

You need to be patient and be able to change horses in midstream. And you need to realize that after your system is in working order and you're reaping the benefits, something may go wrong if you don't operate it properly.

In other words, installing a solar water heater is a do-it-yourselfer's dream come true! Chapters 10 and 12 have loads of information to get you started. Enjoy.

# Literally Going Green with Landscaping

Who cares about the new trend toward "green consciousness?" We're talking a green thumb. Landscaping means you're completely outdoors, and that should be enough to convince any green enthusiast. When you're landscaping, nothing can really break, and you won't run into parts that don't fit. You can start off small, with maybe just a new tree planted in a strategic location, or you can rip out any trees, bushes, or plants you don't like and replace them with ones that support your solar-power habits. Chapter 8 contains all sorts of ideas you can implement.

# Venting Your Attic and Cooling Off the Entire House

A solar attic vent fan project is great because you get to do a little bit of everything (in contrast to the solar water heater project where you get a whole lot of everything). You can install a solar attic vent fan in a day or so, depending on how difficult your attic structure is. You get to install a PV panel, which requires you to figure out orientation and placement. You do some simple wiring without having to be too fussy about quality. You can use electrical tape and just twist the wires together, or you can use a soldering iron and get fancy with shrink tubing. You need to make some drawings of your attic, and really think about prevailing winds and pressures and how they affect air movement. You need to figure out the best spot to install the fan, and you need to know a little about fan technology, but if you don't, it's okay because your system will still work pretty well.

You can't get shocked, no matter how hard you try — a good thing, although some of you are surely disappointed by this.

You may fall off your roof, even if you don't try. I know of at least ten guys who have fallen off roofs. (I would use the gender-neutral term "people," but I don't know any women who have fallen off a roof.) All the do-it-yourselfers I know are envious of these guys, and I suspect they secretly wish they could fall off a roof, too.

Ultimately, a solar attic vent fan works really well. I installed one on my own roof, and the house was much cooler the very next day. We don't have to turn on our air conditioner nearly as much. The problem is, nobody can really see it, so it's hard to brag about. For project instructions, turn to Chapter 13.

# Sheltering Living Spaces from the Sun

Trellises are good projects because they're functional as well as nice looking. You can make them out of wood or synthetic material. You can even get kits of aluminum parts. Plant some flowering vines around your trellis, and the results will be fantastic.

Awnings allow you to really shine in the design department. You can sauce up a plain old window with a good awning, and if you design it right, you can take advantage of the winter sunshine and then block the summer sun from shining in.

Overhangs are really nothing more than glorified awnings. You can really change the look and functionality of a house with one strategically designed overhang. You can take a rectangular, boxy house and make it look interesting and appealing for a very low cost. You don't need a county building permit or inspection if you keep your project reasonably modest.

Chapter 9 has more tips for each of these projects.

# Warming up the Water with an Off-Grid Solar Swimming Pool Heater

You can make a solar hot water heater out of landscape tubing, and you can design it to fit just about anywhere. It's cheap and just as effective as the collectors that cost $600 apiece.

You can connect a small submersible pump to a PV panel, and if you know a few rudimentary things about pumps and pressures and flow, you can make water flow through your homemade collector without having to run it through the pool pump. Solar collectors don't care how fast the water is running through them; they'll put out the same amount of heat with fast or slow flow. So your small system will heat your pool just as much as if you connected it to the pool pump, and it won't load your pump, which makes it cost more to run.

Here's the thing: You want to run your pool pump as little as possible, but you want to get the most amount of heat you can from your solar collectors. So the PV-powered system runs whenever sunshine hits the PV panel. This is ideal because that's when the collectors can heat the water.

If you're on a TOU rate structure, you don't want to run your electric pool pump in the afternoon when rates are high. But this is when you need to run the pump to get heat out of the panels. Your homemade solar pool-heating system solves this problem nicely. Chapter 11 has solar projects, ranging from the simple to the more involved, for your pool.

# Providing an Endless Source of Purified Drinking Water

The water purifier project I describe in Chapter 10 is a good project because it's relatively easy, foolproof, and cheap. Or as cheap as you want to make it. You can make a small purifier or a big one with valves and drippers and all kinds of features. You can purify a gallon of water a day or a thousand, and the concept is the same. You can use salvaged parts, so if your game is to get everything for free, here you go.

This project is good if you've got a remote cabin (off-grid enthusiasts!) and have access to a creek that's of dubious origin (all together now — what do bears do in the woods?). Or if you just want to purify your local tap water, you can use this project for that purpose, too.

# Now You're Really Cooking

Solar cooking is fun and interesting; it teaches you about solar radiation and how it can vary in different weather patterns.

You can build a simple oven, or you can get fancy and make a super-duper one. You can even design and build an *automatic tracker,* which follows the sun over the course of the day automatically, so that you don't have to do it manually. You can make an automatic tracker in several different ways, which all involve interesting physics.

You can make a solar oven any size you want; Chapter 9 has details for building a basic one and a sturdier one. I have a small solar oven next to my barbecue, and I can cook corn on the cob while flipping the ribs. For some reason, the corn has a unique, very pure flavor.

# Charging Your Batteries the Solar Way

This isn't really a project because you don't actually build anything. But using a solar battery charger requires strategy and scheduling, and it gives a good lesson on solar power and PV panels. This is probably one of the easiest ways to start using solar power. Chapter 9 explains what you need to know about getting the most bang for your solar beam.

# Reading Under the Sun at Night

An off-grid reading light is a great project, and you can get as fancy as you want. Start with a simple, self-contained camping light comprised of a small PV panel and a little pod containing a rechargeable battery, a light sensor, a switch, and a few LEDs. Then hang this light over your favorite reading chair (you may need to monkey around to get the light to shine in just the right spot), and you can read every night for hours without plugging in to the grid. Get the lowdown on these lights in Chapter 9.

For those of you who are more ambitious when it comes to electronics, you can design your own circuit (they can be very easy or very clever and efficient). You can design your own optics, which is a real gas if you know a little about the subject. You can get cheap optics parts from Edmund Scientific.

# Pumping Water to New Heights with a Solar Fountain

Solar fountains have tons of potential for do-it-yourselfers. You can make a small one or a massive one. You can make them work any number of ways. You can make one work all night long, if you want.

If your back is strong and your will ironclad, you can use indigenous rocks and mortar to build a cascading waterfall in your backyard that everybody will marvel at. You can make one very noisy with water splashing like Niagara Falls, or you can make a modest, gentle trickle. You can toss some koi into a pond and create a bio system complete with water plants and nightly raccoon raids.

The technology is simple, but you'll need to do some wiring, and you'll need to think about how the system components affect the whole. You need to know water pressure, flow rates, valves, electrical current, and so on (Chapter 8 can get you started). All in all, you'll have to contemplate tons of physics, and that's half the fun.

# Getting Creative with a Solar-Powered Sprinkler

Of all the great toys and gadgets the world of solar offers, the clear winner for most fun is the solar-powered sprinkler. Consider this project a bonus, the 11th installment in a list of ten items.

Here's how it works. You connect the unit (including the sprinkler head and motion detector) to your garden hose and sprinkler and stake it firmly into the ground. Then turn the faucet on. When the motion detector senses movement, the sprinkler discharges a blast of water; then it turns off.

The practical use for a solar-powered sprinkler is to keep animals at bay. For example, you can keep deer from eating all your landscaping, skunks out of the cat food, or raccoons off the roof. When the motion detector senses their presence, it sprays, they skedaddle.

But you can also have some fun with this device. Install it on a hot summer afternoon, and the next time the kids run by, they'll be delighted by the surprise shower. Or place it next to the sidewalk, and when unsuspecting neighbors stroll by on their nightly walk, watch them suddenly pick up the pace to avoid the sprinkles. Just remember: There's no harm in an occasional practical joke, but don't do it too often. You know what they say about paybacks!

# Chapter 25

# Ten Cheap Solar Projects and Devices

*W*hen starting a new hobby (or, maybe in this case, a new lifestyle), it's wise to start out simple and work your way up. If you plan on going for broke and installing a full-scale PV system one day, you should understand the ins and outs of photovoltaics first, because then you'll know how to get the most out of your system. And it's natural to be skeptical of solar power; maybe you want to see what it's all about before you spend a big chunk of your hard-earned money on a major investment.

Personally, I like small projects. Get in, get out, and move on. That's what you find in this chapter. And if you've got kids, some of these projects are ideal to do together with them because they're easy and fun. And maybe you can even sneak in a science lesson or two. (You can find more information on most of these ideas in Chapter 9.)

## Landscape Lighting

This is probably the cheapest way to get started using solar power. You can install a single light in a strategic location and get a lot of effect. It takes five minutes and costs less than $10. You won't get a jolt of electricity, and if the lights don't illuminate an area exactly the way you had planned, you can move them around to your heart's content.

One of my all-time favorite gizmos is a swimming pool light that changes colors. It floats on the surface and casts its light down into the pool so the entire pool changes colors. The lights are fun to swim underneath, and kids love to play games with them. (A word of caution, though: Kids usually end up breaking them.)

The same color-changing scheme is used to good effect in some decorative yard lights. They add just a hint of mystique to your yard and gently catch the eye. If you have a big picture window, put one of these outside it.

## Portable Showers

Portable showers are kind of silly, I'll grant that. You fill a big plastic bag with water and set it in the sun. When the water heats up enough (to 105°F or so), you hang the bag from a tree, stand underneath it, open a small plastic valve, and voilà, you have a shower au naturel. Ahhh . . .

You can use one near your swimming pool to rinse the chlorine off after a swim. You can use them while camping. You can use one for a hot water bottle big enough to soothe your entire back or stomach.

## Sunscreens with Roll-up Mechanisms

A lot of sunlight comes into your non-northern windows over the course of a hot day. This has two negative effects. First, the sunlight makes the room temperature hotter, and second, the bright light makes the room seem hotter (a psychological effect). It's a fact that if two rooms are the same exact temperature, but one has a lot of sunlight and the other is dark, people prefer the darker room.

Solar screens reduce the amount of sunlight coming into a room by up to 90 percent. It costs around $2 per year to cover a typical window, and the effect is dramatic. You get the most heat reduction in your house when you install sunscreens on the biggest south-facing windows.

You can tack sunscreens up with ordinary thumbtacks, and they look good from both inside and out if you take care to cut the screen corners flush with the window frame. You can make a retractor mechanism for a few dollars. This device allows you to retract the screen when you don't want it to cover the window. (For the same reason a room seems cooler in the summer with a screen, the room will seem warmer in the winter without the screen.)

# Solar Fountains

You can get a good solar fountain for under $100. The water will flow when the sky is sunny and cut back when it's cloudy. You'll be aware of how much sunshine you are receiving at any given time.

You can build a large solar fountain for a little bit more if you do most of the labor yourself and use natural landscaping elements, like indigenous rocks. You can build a really big fountain, and the only cost you'll incur is for the solar pump.

You can build a solar fountain with a large upper reservoir so you can even out the fluctuations in water flow when the sun changes. The reservoir also allows you to save the water for use later, like when you're dining nearby.

You can locate a solar fountain anywhere, as long as you put the PV panel in sunlight.

# Sun Tea

Here's the cheapest project you can possibly do with the sun. Put some tea bags and water in a big glass jar with a lid, and leave it in the sunshine for a few hours. By the middle of the afternoon, you'll have refreshing sun tea. You don't need to heat water on the stove or in the microwave, and the tea tastes wonderful. (Just don't leave the jar in the sun too long.)

# Battery Chargers

If you use a lot of batteries, you can drastically cut back on battery costs by using rechargeable batteries in conjunction with a solar charger. And I'm not only talking about the standard household batteries, either. You can get solar chargers for vehicle batteries, notebook computer batteries, and small appliances.

Using these devices gives you a very good lesson on PV technology because you'll need to locate the PV panel in a good, sunny spot to get the most juice into your batteries.

# Solar Cooking

Here's a fun and potentially productive project that anybody can do. For less than $50, you can make a good solar oven. You can easily save this much in power bills over the course of a sunny season, and if you make your oven with enough quality, you can use it most of the year. Plus, in the summer, you won't be heating your house with conventional cooking.

# Solar Hat Fans

This is a cheap toy, but a fun one. Give solar hat fans as gifts that people will remember for their novelty. Kids will have a gas with them. Solar hat fans are small solar-powered fans that clip onto your hat visor, and they cost only $10 apiece. If you've got a business, put your logo on the fans and hand them out to potential customers. Look up www.realgoods.com for a supply.

# Solar Lamps

For around $40, you can get a little light module that you can charge during sunlight hours and use when it gets dark. They're similar to solar landscape lighting and just as effective. You can read with the light or hang one in your tent for a cool light that you can play cards under for hours.

# Solar Flashlights

Are you always getting frustrated when you reach for your flashlight and its batteries are pooped out again? A solar flashlight is always ready to go when you need it. Instead of storing it in a drawer, you need to put it on a window sill, but it's worth it. Kids love these lights, and they're much brighter than you think. They work by using well-designed optics and efficient LEDs. They're great for remote cabins, boats, camping, RVs, and so on. In contrast to many "cheap" solar gizmos, these are reliable and provide you with light for years, so they're also good investments.

# Chapter 26

# Ten or So Worst Solar Mistakes

**B**elieve it or not, you can abuse solar power. I don't mean just by laying out naked in the sun for hours without sunscreen, but rather by installing ill-advised projects or not attending to the details as well as you should. For some of you, this chapter is a reminder of what not to do. For others, it's a how-to seminar.

## Installing Like Curly, Moe, and Larry

I watched a Three Stooges movie where Curly plugged a leak in a shower by connecting a pipe to it. Then the other end of the pipe leaked, so he connected another pipe to that. The other end of that pipe leaked, so he installed another pipe. Pretty soon, the bathroom was filled with pipes. Moe came in and slapped Curly and then pulled Larry's hair and slapped him, too. Curly retaliated by whacking Moe on the head with a large wrench, and so it went.

If you fail to see the humor in this, you probably should not install a solar water heater. In fact, anytime you install anything, keep your sense of humor handy.

## Opting for a Solar Swimming Pool Heater Rather Than a Solar Cover

Wait a minute, you're thinking. This project was listed in the ten best solar projects chapter!

Exactly. Solar pool heaters are very expensive, and you don't get any return on investment from them. They don't last that long because sunshine and chlorine eat them up. They're hard to maintain because they're on your roof, and they leak everywhere when they break (not *if* they break, *when* they break).

And you know what? A solar pool cover is about 2 percent of the cost and works better. Plus, if you've got any sort of respect for Mother Nature, you don't mind freezing cold water because that's exactly the way she serves it up most of the time. So what's a little cold water? Instead of getting out of bed in the morning and flipping on your coffee pot, jump into your freezing pool. You'll be awake in no time (and think of all the energy you'll save).

# Going Out of Your Mind by Going Off-Grid

If you have a choice between going off-grid and installing a PV intertie system, do *not* go off-grid. The economics are horrible, in a relative sense. Off-grid systems use batteries, which are operational and waste nightmares. Plus, you can't get a mortgage for an off-grid house.

Nevertheless, a movement is afoot to "go back to Mother Nature" by going off-grid. This movement would be better titled, "getting away from humanity while punching Mother Nature in the gut."

You can save a heckuva lot more pollution by going intertie than off-grid in many ways. If you want to get away from humanity (I'm empathetic, believe me), go backpacking and remember to take biodegradable toilet paper. Then when you get back home, check your intertie PV system to see how much of a net surplus of energy you generated.

# Relying on Solar When You Rarely See the Sun

If you're in a cloudy, rainy, cold climate, and your utility rates are low, don't install a PV system, no matter how much you want to jump on the bandwagon. You can use your money to save the world in many other planet-friendly, more efficient ways. Install a geothermal system for heating and cooling. Put in a pellet stove and figure out how to work it perfectly. Give to Greenpeace.

# Cheaping Out on Water Heater Systems

If you review the availability of solar water heater systems, you'll find a large range of prices and performances, and you'll discover that some systems don't work very well in cold climates. If you're in a climate where temperatures can drop below freezing, the pipes or the collector in some solar water heaters can burst (but the manufacturer will never tell you that).

Now picture this scenario: It took six men to lift the collector onto your roof, and this was after it took an hour to pull the thing out of its very large, very expensive packing crate that was delivered via truck because UPS doesn't deliver anything that big. So now you need to get six men to lift it back down, repack it, and send it back to the factory for warranty repairs. The company fixes it, repacks it, then trucks it back to you (warranty repairs generally don't cover shipping costs, so you'll have to pay those). Then you must find six men to lift the collector back into place on your roof. But you still have the same problem: You cheaped out and bought a system not made for your particular climate.

The point of all this? Be very thorough when researching and selecting a solar water heater system. It's the best thing under the sun, but it can be the worst if you don't do your homework.

# Your Eyes Are Bigger Than Your Stomach

Any solar system that generates more energy than you use is a waste. You save nothing by using a solar system at less than its full capacity. If you install a solar water heater that's too big, you've wasted money. If you install an intertie system that's too big, you get nothing for the excess power that you don't use (although to be fair, this is changing).

# Skipping the Groundwork

Strictly speaking, you don't have to do an energy audit and take conservation steps before installing a solar system. But bypassing these processes violates the spirit of solar philosophy. Actually, I don't really have to put this item in this list because you're already an environmentalist, right?

# Going with the Cheapest Bid Because It's the Cheapest

Do I really need to elaborate? In case I do, let me just remind you that you get what you pay for. Cheap usually means more headaches, shoddy quality, and more money in the long run. Spend a little more money up front for quality equipment and better workmanship. In the end, you'll be glad you did.

# Ignoring Murphy's Law

If you think I'm just joking when I talk about Murphy's Law, think again. I've been doing projects since I was 5. I've done easy ones, I've done hard ones. I've been in the linear particle accelerator at Stanford University working on projects with doctors and engineers who have fancy degrees and thousands of years' worth of experience. Here's the bottom line: Nothing ever goes the way it's supposed to. Never (unless your goal is to do something stupid; then things tend to work out even better than you planned).

Measure twice, cut once. Then be prepared for anything, and always be patient. Every project involves art and science. The science part is usually obvious, and that's what attracts most people. The art is in smiling when things go awry.

# Index

## Business/Accounting & Bookkeeping

Bookkeeping For Dummies
978-0-7645-9848-7

eBay Business
All-in-One For Dummies,
2nd Edition
978-0-470-38536-4

Job Interviews
For Dummies,
3rd Edition
978-0-470-17748-8

Resumes For Dummies,
5th Edition
978-0-470-08037-5

Stock Investing
For Dummies,
3rd Edition
978-0-470-40114-9

Successful Time
Management
For Dummies
978-0-470-29034-7

## Computer Hardware

BlackBerry For Dummies,
3rd Edition
978-0-470-45762-7

Computers For Seniors
For Dummies
978-0-470-24055-7

iPhone For Dummies,
2nd Edition
978-0-470-42342-4

Laptops For Dummies,
3rd Edition
978-0-470-27759-1

Macs For Dummies,
10th Edition
978-0-470-27817-8

## Cooking & Entertaining

Cooking Basics
For Dummies,
3rd Edition
978-0-7645-7206-7

Wine For Dummies,
4th Edition
978-0-470-04579-4

## Diet & Nutrition

Dieting For Dummies,
2nd Edition
978-0-7645-4149-0

Nutrition For Dummies,
4th Edition
978-0-471-79868-2

Weight Training
For Dummies,
3rd Edition
978-0-471-76845-6

## Digital Photography

Digital Photography
For Dummies,
6th Edition
978-0-470-25074-7

Photoshop Elements 7
For Dummies
978-0-470-39700-8

## Gardening

Gardening Basics
For Dummies
978-0-470-03749-2

Organic Gardening
For Dummies,
2nd Edition
978-0-470-43067-5

## Green/Sustainable

Green Building
& Remodeling
For Dummies
978-0-470-17559-0

Green Cleaning
For Dummies
978-0-470-39106-8

Green IT For Dummies
978-0-470-38688-0

## Health

Diabetes For Dummies,
3rd Edition
978-0-470-27086-8

Food Allergies
For Dummies
978-0-470-09584-3

Living Gluten-Free
For Dummies
978-0-471-77383-2

## Hobbies/General

Chess For Dummies,
2nd Edition
978-0-7645-8404-6

Drawing For Dummies
978-0-7645-5476-6

Knitting For Dummies,
2nd Edition
978-0-470-28747-7

Organizing For Dummies
978-0-7645-5300-4

SuDoku For Dummies
978-0-470-01892-7

## Home Improvement

Energy Efficient Homes
For Dummies
978-0-470-37602-7

Home Theater
For Dummies,
3rd Edition
978-0-470-41189-6

Living the Country Lifestyle
All-in-One For Dummies
978-0-470-43061-3

Solar Power Your Home
For Dummies
978-0-470-17569-9

## Internet

Blogging For Dummies,
2nd Edition
978-0-470-23017-6

eBay For Dummies,
6th Edition
978-0-470-49741-8

Facebook For Dummies
978-0-470-26273-3

Google Blogger
For Dummies
978-0-470-40742-4

Web Marketing
For Dummies,
2nd Edition
978-0-470-37181-7

WordPress For Dummies,
2nd Edition
978-0-470-40296-2

## Language & Foreign Language

French For Dummies
978-0-7645-5193-2

Italian Phrases
For Dummies
978-0-7645-7203-6

Spanish For Dummies
978-0-7645-5194-9

Spanish For Dummies,
Audio Set
978-0-470-09585-0

## Macintosh

Mac OS X Snow Leopard
For Dummies
978-0-470-43543-4

## Math & Science

Algebra I For Dummies
978-0-7645-5325-7

Biology For Dummies
978-0-7645-5326-4

Calculus For Dummies
978-0-7645-2498-1

Chemistry For Dummies
978-0-7645-5430-8

## Microsoft Office

Excel 2007 For Dummies
978-0-470-03737-9

Office 2007 All-in-One
Desk Reference
For Dummies
978-0-471-78279-7

## Music

Guitar For Dummies,
2nd Edition
978-0-7645-9904-0

iPod & iTunes
For Dummies,
6th Edition
978-0-470-39062-7

Piano Exercises
For Dummies
978-0-470-38765-8

## Parenting & Education

Parenting For Dummies,
2nd Edition
978-0-7645-5418-6

Type 1 Diabetes
For Dummies
978-0-470-17811-9

## Pets

Cats For Dummies,
2nd Edition
978-0-7645-5275-5

Dog Training For Dummies,
2nd Edition
978-0-7645-8418-3

Puppies For Dummies,
2nd Edition
978-0-470-03717-1

## Religion & Inspiration

The Bible For Dummies
978-0-7645-5296-0

Catholicism For Dummies
978-0-7645-5391-2

Women in the Bible
For Dummies
978-0-7645-8475-6

## Self-Help & Relationship

Anger Management
For Dummies
978-0-470-03715-7

Overcoming Anxiety
For Dummies
978-0-7645-5447-6

## Sports

Baseball For Dummies,
3rd Edition
978-0-7645-7537-2

Basketball For Dummies,
2nd Edition
978-0-7645-5248-9

Golf For Dummies,
3rd Edition
978-0-471-76871-5

## Web Development

Web Design All-in-One
For Dummies
978-0-470-41796-6

## Windows Vista

Windows Vista
For Dummies
978-0-471-75421-3